T0262726

Automatic Flight Control Systems

Part 1

Literature Review and Theoretical Developments

Automatic Flight Control Systems

Edited by **Margaret Ziegler**

Clanrye International

New Jersey

Published by Clanrye International,
55 Van Reypen Street,
Jersey City, NJ 07306, USA
www.clanryeinternational.com

Automatic Flight Control Systems
Edited by Margaret Ziegler

© 2015 Clanrye International

International Standard Book Number: 978-1-63240-073-4 (Hardback)

This book contains information obtained from authentic and highly regarded sources. Copyright for all individual chapters remain with the respective authors as indicated. A wide variety of references are listed. Permission and sources are indicated; for detailed attributions, please refer to the permissions page. Reasonable efforts have been made to publish reliable data and information, but the authors, editors and publisher cannot assume any responsibility for the validity of all materials or the consequences of their use.

The publisher's policy is to use permanent paper from mills that operate a sustainable forestry policy. Furthermore, the publisher ensures that the text paper and cover boards used have met acceptable environmental accreditation standards.

Trademark Notice: Registered trademark of products or corporate names are used only for explanation and identification without intent to infringe.

Printed in the United States of America.

Contents

Preface

The purpose of the book is to provide a glimpse into the dynamics and to present opinions and studies of some of the scientists engaged in the development of new ideas in the field from very different standpoints. This book will prove useful to students and researchers owing to its high content quality.

The history of flight control cannot be considered separately to the history of aviation. Since the early days, the conception of automatic flight control systems has advanced from mechanical control systems to greatly developed automatic fly-by-wire flight control systems which can be found in military jets and civil airliners these days. Even today, several research attempts are made for the further advancement of these flight control systems in numerous aspects. Current advancements in this area target a variety of different aspects. This book presents a collection of knowledge on important research areas, like inertial navigation, handling of unmanned airplanes and helicopters, trajectory control of an unmanned space re-entry automobile, aeroservoelastic control, modifying flight control, and error tolerant flight control. It discusses theoretical outlook and current conceptual advancements in flight control systems along with describing theories of modified and fault-tolerant flight control systems. Each technique has been elaborated using illustrations and appropriate examples.

At the end, I would like to appreciate all the efforts made by the authors in completing their chapters professionally. I express my deepest gratitude to all of them for contributing to this book by sharing their valuable works. A special thanks to my family and friends for their constant support in this journey.

Editor

1

Fundamentals of GNSS-Aided Inertial Navigation

Ahmed Mohamed and Apostolos Mamatas
University of Florida
USA

1. Introduction

GNSS-aided inertial navigation is a core technology in aerospace applications from military to civilian. It is the product of a confluence of disciplines, from those in engineering to the geodetic sciences and it requires a familiarity with numerous concepts within each field in order for its application to be understood and used effectively. Aided inertial navigation systems require the use of kinematic, dynamic and stochastic modeling, combined with optimal estimation techniques to ascertain a vehicle's navigation state (position, velocity and attitude). Moreover, these models are employed within different frames of reference, depending on the application. The goal of this chapter is to familiarize the reader with the relevant fundamental concepts.

2. Background

2.1 Modeling motion

The goal of a navigation system is to determine the state of the vehicle's trajectory in space relevant to guidance and control. These are namely its position, velocity and attitude at any time. In inertial navigation, a vehicle's path is modeled kinematically rather than dynamically, as the full relationship of forces acting on the body to its motion is quite complex. The kinematic model incorporates accelerations and turn rates from an inertial measurement unit (IMU) and accounts for effects on the measurements of the reference frame in which the model is formalized. The kinematic model relies solely on measurements and known physical properties of the reference frame, without regard to vehicle dynamic characteristics. On the other hand, in incorporating aiding systems like GNSS, a dynamic model is used to predict error states in the navigation parameters which are rendered observable through the external measurements of position and velocity. The dynamics model is therefore one in which the errors are related to the current navigation state. As will be shown, some errors are bounded while others are not. At this point, we make the distinction between the aided INS and free-navigating INS. Navigation using the latter method represents a form of "dead reckoning", that is the navigation parameters are derived through the integration of measurements from some defined initial state. For instance, given a measured linear acceleration, integration of the measurement leads to velocity and double integration results in the vehicle's position. Inertial sensors exhibit biases and noise that, when integrated, leads to computed positional drift over time. The goal of the aiding system is therefore to help estimate the errors and correct them.

2.2 Reference frames

Proceeding from the sensor stratum up to more intuitively accessible reference systems, we define the following reference frames:

- Sensor Frame (*s*-frame). This is the reference system in which the inertial sensors operate. It is a frame of reference with a right-handed Cartesian coordinate system whose origin is at the center of the instrument cluster, with arbitrarily assigned principle axes as shown in figure 1.

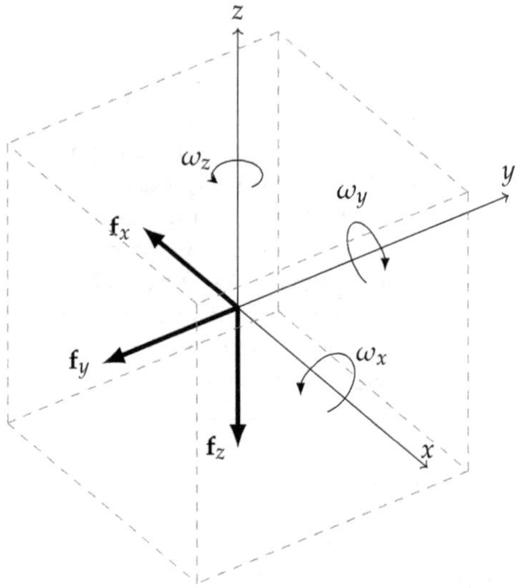

Fig. 1. IMU measurements in the *s*-frame

- Body Frame (*b*-frame). This is the reference system of the vehicle whose motions are of interest. The *b*-frame is related to the *s*-frame through a rigid transformation (rotation and translation). This accounts for misalignment between the sensitive axes of the IMU and the primary axes of the vehicle which define roll, pitch and yaw. Two primary axis definitions are generally employed: one with $+y$ pointing toward the front of the vehicle ($+z$ pointing up), and the other with $+x$ pointing toward the nose ($+z$ pointing down). The latter is a common aerospace convention used to define heading as a clockwise rotation in a right-handed system (Rogers, 2003).
- Inertial Frame (*i*-frame). This is the canonical inertial frame for an object near the surface of the earth. It is a non-rotating, non-accelerating frame of reference with a Cartesian coordinate system whose x axis is aligned with the mean vernal equinox and whose z axis is coaxial with the spin axis of the earth. The y-axis completes the orthogonal basis and the system's origin is located at the center of mass of the earth.
- Earth-Fixed Frame (*e*-frame). With some subtle differences that we shall overlook, this system's z axis is defined the same way as for the *i*-frame, but the x axis now points toward the mean Greenwich meridian, with y completing the right-handed system. The origin is at the earth's center of mass. This frame rotates with respect to the *i*-frame at the earth's rotation rate of approximately 15 degrees per hour.

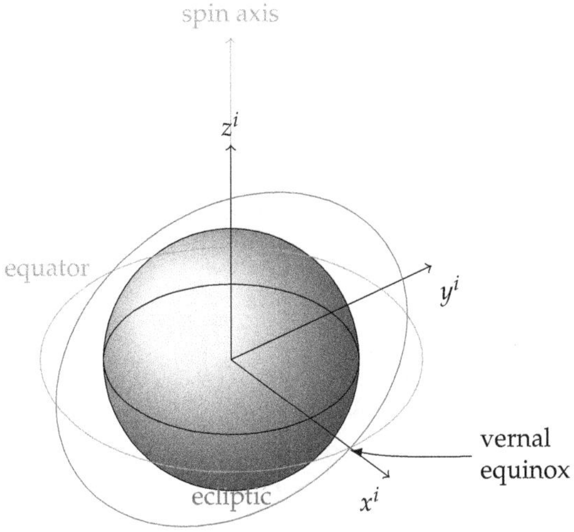

Fig. 2. Inertial Frame

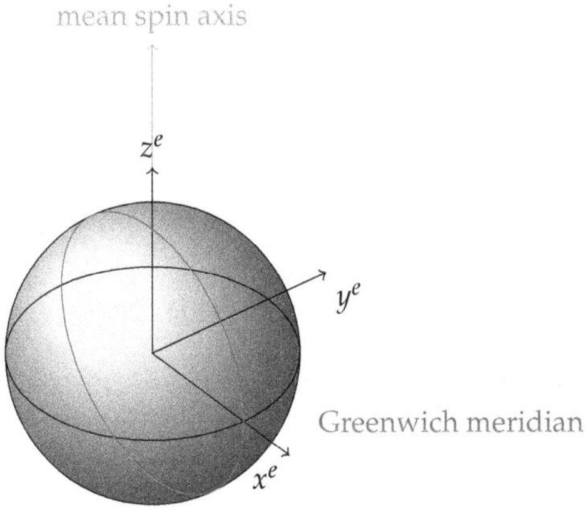

Fig. 3. Earth-Fixed Frame

- Local-Level Frame (l-frame). This frame is defined by a plane locally tangent to the surface of the earth at the position of the vehicle. This implies a constant direction for gravity (straight down). The coordinate system used is *easting, northing, up* (enu), where Up is the normal vector of the plane, North points toward the spin axis of Earth on the plane and East completes the orthogonal system.

Finally, we remark that the implementation of the INS can be freely chosen to be formulated in any of the last three frames, and it is common to refer to the navigation frame (n-frame) once it is defined as being either the i-, e- or l-frames, especially when one must make the distinction between native INS output and transformed values in another frame.

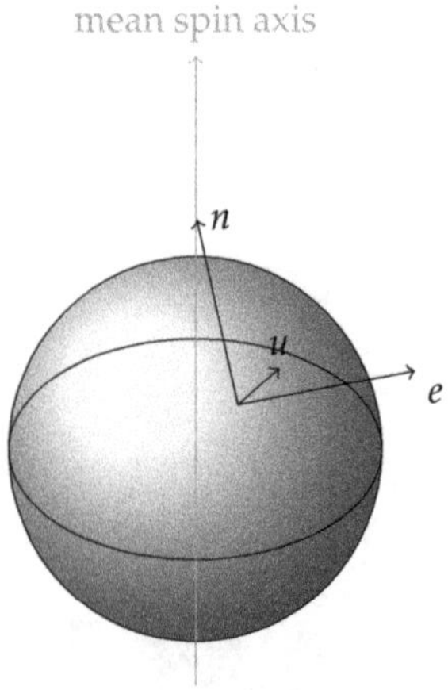

Fig. 4. Local-Level Frame

2.3 Geometric figure of the earth

Having defined the common reference frames, we must consider the size and shape of the earth itself, an especially important topic when moving between the *l*- and *e*- frames or when converting Cartesian to geodetic (latitude, longitude, height) coordinates. The earth, though commonly imagined as a sphere, is in fact more accurately described as an ellipse revolved around its semi-major axis, an *ellipsoid*. Reference ellipsoids are generally defined by the magnitude of their semi-major axis (equatorial radius) and their flattening, which is the ratio of the polar radius to the equatorial radius. Since the discovery of the elipticity of the earth, many ellipsoids have been formulated, but today the most important one for global navigation is the WGS84 ellipsoid[1], which forms the basis of the WGS84 system to which all GPS measurements and computations are tied (Hofmann-Wellenhof et al., 2001). The WGS84 ellipsoid is defined as having an equatorial radius of 6,378,137 m and a flattening of 1/298.257223563 centered at the earth's center of mass with 0 degrees longitude located 5.31 arc seconds east of the Greenwich meridian (NIMA, 2000; Rogers, 2003). It is worth defining another ellipsoidal parameter, the eccentricity e, as the distance of the ellipse focus from the axes center, and is calculated as

$$e^2 = \frac{a^2 - b^2}{a^2} \tag{1}$$

Figure 5 shows a cross-sectional view of the reference ellipsoid with having semi-major and semi-minor dimensions a and b, respectively. Note that b is derivable from a and f. A point P is located at height h normal to the surface. N is the radius of curvature in the prime vertical of the ellipsoid at this point[2]. The angle between the x, y plane and the surface normal vector of P is the geodetic latitude ϕ. Note that the loci of normal vectors that pass through the centroid of the ellipsoid are constrained to the equator and the meridians. This means that, in general, the geodetic latitude ϕ is not the same as the *geocentric* latitude ψ, as shown in figure 6. The

[1] A variant of the GRS80 ellipsoid

[2] This is also called the normal radius of curvature, hence the symbol N.

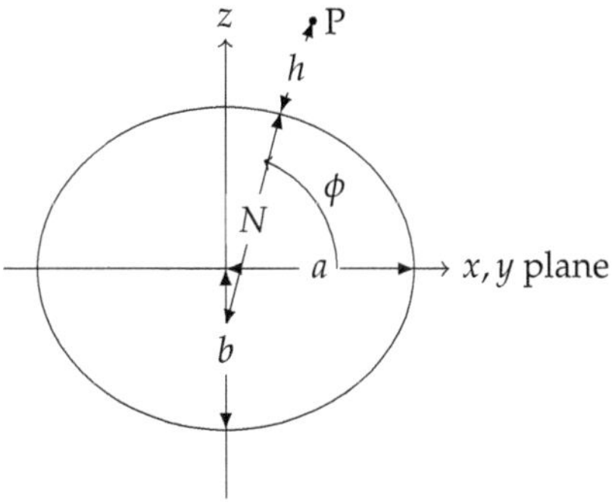

Fig. 5. Reference Ellipsoid

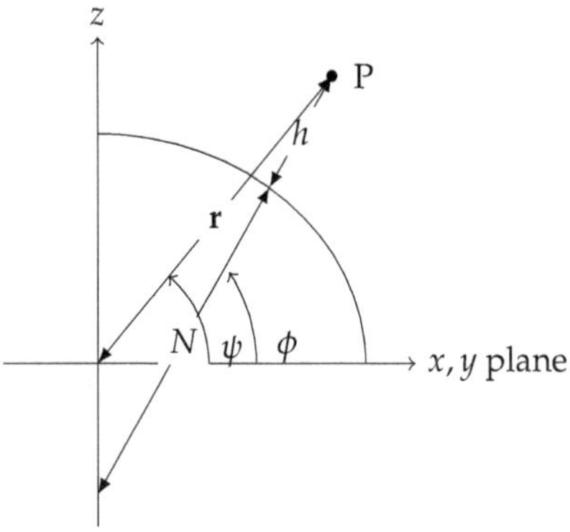

Fig. 6. Geodetic vs. Geocentric Latitude

value of N is obtained by

$$N = \frac{a}{\sqrt{(1 - e^2 \sin^2 \phi)}} \tag{2}$$

Another radius is defined, namely the radius of curvature in the meridian, M, which is given as

$$M = \frac{a(1 - e^2)}{(1 - e^2 \sin^2 \phi)^{3/2}} \tag{3}$$

The two parameters N and M are necessary for calculating the linear distances and velocity components from the geodetic coordinate system in the local-level frame. In order to relate geodetic position changes and linear distances, we begin with the simple case of a sphere of radius R_e. Note that the linear distance between two points along a meridian (in the North direction) is

$$\delta n = (R_e + h)\delta\psi \tag{4}$$

and the distance along a parallel (in the East direction) is

$$\delta e = (R_e + h) \cos \psi \delta \lambda \tag{5}$$

where h is the height above the sphere. In the case of the ellipsoidal earth, one radius does not suffice to reduce both directions of motion to linear distances and the equations become

$$\delta n = (M + h)\delta \phi \tag{6}$$

$$\delta e = (N + h) \cos \phi \delta \lambda \tag{7}$$

2.4 Gravitation and gravity

Inertial navigation relies on measurements made in an inertial reference frame, *i.e.* one free of acceleration or rotation. Vehicles near the earth's surface, of course, are subjected to both of these factors. As an accelerometer is not capable of distinguishing accelerations due to motion and accelerations arising from reaction forces in a gravity field, we must have *a priori* knowledge of the earth's gravitation in order to subtract its effects from sensor measurements. The gravitational field of the earth is described by its potential V at a point $\mathbf{P}$ such that

$$\mathbf{V}(\mathbf{P}) = G \iiint_{earth} \frac{\rho(\mathbf{Q})}{l} dv_{\mathbf{Q}} \tag{8}$$

where $\mathbf{Q}$ is a point within the earth with mass density $\rho(\mathbf{Q})$ and volume element $dv_{\mathbf{Q}}$, located at a distance l from $\mathbf{P}$ and G is the gravitational constant (Hofmann-Wellenhof & Moritz, 2005). The gravitational vector field is defined as the gradient of the potential:

$$\bar{\mathbf{g}}^e = \nabla \mathbf{V}^e = \frac{\partial \mathbf{V}^e}{\partial \mathbf{r}^e} = \begin{pmatrix} V_x \\ V_y \\ V_z \end{pmatrix} \tag{9}$$

where $\bar{\mathbf{g}}^{\mathbf{e}}$ is the gravitational vector associated with the position vector $\mathbf{r}^{\mathbf{e}}$. In practice, sensor measurements are further conditioned by the rotation of the earth. The difference of gravitation and centripetal acceleration caused by Earth's rotation is embodied in the *gravity vector* and is more commonly used in practice. It is defined as

$$\mathbf{g}^e = \bar{\mathbf{g}}^e - \mathbf{\Omega}_e \mathbf{\Omega}_e \mathbf{r}^e \tag{10}$$

where Ω_{ie} is the skew-symmetric representation of the earth's rotation rate (the skew-symmetric matrix representation is treated in section 2.6). In the local-level frame, the gravity vector is expressed as

$$\mathbf{g}^l = \begin{pmatrix} 0 \\ 0 \\ -(\gamma + \sigma g_u) \end{pmatrix} \tag{11}$$

where γ is normal gravity and σg_u is a disturbance in the vertical component of the vector (Hofmann-Wellenhof & Moritz, 2005).[3]

[3] We leave out the non-vertical deflections here for brevity, but a more detailed treatment should include them.

2.5 Normal gravity

Uneven mass distributions within the earth, as well as departure of its actual shape from a perfect ellipsoid leads to a highly complex gravity field. It is therefore convenient for navigation purposes to approximate the gravity field using the so-called *normal gravity model,* computed in closed form using Somigliana's formula (Schwarz & Wei, 1990; Torge, 2001) :

$$\gamma_0 = \gamma_a \frac{1 + k \sin^2 \phi}{\sqrt{1 - e^2 \sin^2 \phi}} \tag{12}$$

where

$$k = \frac{b\gamma_b}{a\gamma_a} - 1$$

$$\gamma_a, \gamma_b = \text{equatorial and polar gravity values}$$

A computationally faster method to calculate gravity involves expanding (12) by power series with respect to e^2 and truncating, yielding:

$$\gamma_0 = \gamma_a(1 + \beta \sin^2 \phi + \beta_1 sin^2 2\phi) \tag{13}$$

where γ_a is the gravity at the equator, β is the "gravity flattening" term (Hofmann-Wellenhof & Moritz, 2005), defined as

$$\beta = \frac{\gamma_b - \gamma_a}{\gamma_a} \tag{14}$$

The second parameter in (13) β_1 is given by

$$\beta_1 = \frac{f^2}{4} - \frac{5f\omega_e^2 a}{8\gamma_a} \tag{15}$$

where ω_e is the earth rotation constant. This approximation of (12) is accurate to approximately 0.1 μg (Featherstone & Dentith, 1997), which is sufficient for navigation purposes. Higher-accuracy approximations are given in (Hofmann-Wellenhof & Moritz, 2005; Torge, 2001).Table 1 gives the relevant WGS84 parameters for computing normal gravity. The first four are the defining parameters of the system, while the last two are derived for use in computing normal gravity. Incorporating height h allows a more general formula for gravity

a	6,378,137 m
f	1/298.257223563
ω_e	7.292115×10^{-5} rad/s
GM	$3.986004.418\times10^{14}$m^3/s^2
γ_a	9.7803253359 m/s^2
γ_b	9.8321849378 m/s^2

Table 1. WGS84 parameters

away from the ellipsoidal surface:

$$\gamma = \gamma_0 - (3.0877 \times 10^{-6} - 4.4 \times 10^{-9} \sin^2 \phi)h + 0.72 \times 10^{-12}h^2 \tag{16}$$

2.6 Mathematical treatment of rotations

2.6.1 Direction cosines matrix

Before proceeding to linear and rotational models of motion, we must first discuss the formulation of the rigid-body transformations required to express vectors defined in a particular frame in terms of another frame. These are comprised of translations and rotations, the former being the straightforward operation of addition. We shall therefore direct our attention to rotations and their time-derivatives. In general a rotation matrix is an operator transforming vectors from one orthogonal basis to another. Let $\begin{pmatrix} x & y & z \end{pmatrix}^T$ be a vector $\mathbf{p}^a$ in some frame a and $\begin{pmatrix} x' & y' & z' \end{pmatrix}^T$ be a vector $\mathbf{p}^b$ in frame b, then

$$\begin{aligned}\mathbf{p}^b &= \mathbf{R}_a^b \mathbf{p}^a \\ &= \begin{pmatrix} \mathbf{i}_a \cdot \mathbf{i}_b & \mathbf{j}_a \cdot \mathbf{i}_b & \mathbf{k}_a \cdot \mathbf{i}_b \\ \mathbf{i}_a \cdot \mathbf{j}_b & \mathbf{j}_a \cdot \mathbf{j}_b & \mathbf{k}_a \cdot \mathbf{j}_b \\ \mathbf{i}_a \cdot \mathbf{k}_b & \mathbf{j}_a \cdot \mathbf{k}_b & \mathbf{k}_a \cdot \mathbf{k}_b \end{pmatrix} \mathbf{p}^a \end{aligned} \tag{17}$$

where $\begin{pmatrix} \mathbf{i}_a & \mathbf{j}_a & \mathbf{k}_a \end{pmatrix}$ and $\begin{pmatrix} \mathbf{i}_b & \mathbf{j}_b & \mathbf{k}_b \end{pmatrix}$ are the orthonormal bases of a and b, respectively. Note the use of superscripts to indicate the reference frame. The rotation matrix notation indicates a transformation from the a-frame to the b-frame. Because the basis vectors are of unit length the dot products in $\mathbf{R}_a^b$ define the cosines of the angles between the vector pairs, therefore the rotation matrix is also commonly known as the *direction cosines matrix* (DCM). The two properties of DCMs in a right-handed Cartesian system are:

1. $(\mathbf{R}_a^b)^{-1} = (\mathbf{R}_a^b)^T = \mathbf{R}_b^a$
2. $\det(\mathbf{R}_a^b) = 1$

DCMs in $\mathbb{R}^3$ are decomposable into three elemental rotations performed sequentially about a principal axis in the originating frame, thence about the rotated and twice-rotated remaining axes(Kuipers, 1999). The elemental rotations about the x, y and z axes are defined as

$$\mathbf{R}_x(\varphi) = \begin{pmatrix} 1 & 0 & 0 \\ 0 & \cos\varphi & \sin\varphi \\ 0 & -\sin\varphi & \cos\varphi \end{pmatrix}, \mathbf{R}_y(\theta) = \begin{pmatrix} \cos\theta & 0 & -\sin\theta \\ 0 & 1 & 0 \\ \sin\theta & 0 & \cos\theta \end{pmatrix}, \mathbf{R}_z(\psi) = \begin{pmatrix} \cos\psi & \sin\psi & 0 \\ -\sin\psi & \cos\psi & 0 \\ 0 & 0 & 1 \end{pmatrix} \tag{18}$$

The choice of axis order is mathematically arbitrary, but while working in the ENU definition of the l-frame we must employ a y, x, z sequence when defining the rotation from the mechanization frame to the body frame. The transformation from the body frame to the navigation frame is therefore composed of the inverse (Titterton & Weston, 2004) , that is

$$\mathbf{R}_b^n = \mathbf{R}_n^{bT} = \mathbf{R}_z(-\psi)\mathbf{R}_x(-\theta)\mathbf{R}_y(-\varphi) \tag{19}$$

where n is any of the valid mechanization frames given in section 2.2. More explicitly, the DCM in terms of the Euler angles is

$$\mathbf{R}_b^n = \begin{pmatrix} \cos\psi\cos\varphi - \sin\psi\sin\theta\sin\varphi & -\sin\psi\cos\theta & \cos\psi\sin\varphi + \sin\psi\sin\theta\cos\varphi \\ \sin\psi\cos\varphi + \cos\psi\sin\theta\sin\varphi & \cos\psi\cos\theta & \sin\psi\sin\varphi - \cos\psi\sin\theta\cos\varphi \\ -\cos\theta\sin\varphi & \sin\theta & \cos\theta\cos\varphi \end{pmatrix} \tag{20}$$

The sequential angular rotations φ, θ, ψ are known as *Euler angles*; if $n = l$, the Euler angles are called roll, pitch, and yaw. Given a DCM, there is no unique decomposition into Euler angles without prior knowledge of the convention. For example, an equally valid DCM could be constructed from the sequence z, x, z or any of a number of permutations (Pio, 1966), but we remind the reader that unless the sequence is defined uniformly for the INS mechanization, the retrieval of heading, roll and pitch angles from a computed DCM may well be meaningless. We therefore stress the order given in (19) and will employ it exclusively moving forward. This being the case, we recover roll, pitch and yaw by

$$(\varphi \; \theta \; \psi)^T = \left(\tan^{-1}\left(-\frac{R_{31}}{R_{33}}\right) \; \sin^{-1}(R_{32}) \; \tan^{-1}\left(-\frac{R_{12}}{R_{22}}\right)\right)^T \tag{21}$$

The representation of rotations as discussed up to now is tied to the historical simplicity of relating measurements of gimballed IMU axis encoders to the DCM. The careful reader will note however that singularities exist in this method. For instance, when a vehicle is pitched up 90 degrees, two axes respond to the same motion and a degree of freedom is lost, leaving no unique roll and heading values that will satisfy the DCM terms. In strap-down systems, this is mathematically equivalent to *gimbal lock* in gimballed INS. Mechanical and algorithmic solutions exist to the problem, but are beyond the scope of this writing. An alternative representation of rotations that does not suffer this problem is therefore sometimes used employing quaternions.

2.6.2 Quaternions

Quaternions are a four-dimensional extension of complex numbers having the form

$$\mathbf{q} = a + b\mathbf{i} + c\mathbf{j} + d\mathbf{k} \tag{22}$$

where a is the real component and b, c and d are imaginary. Quaternion multiplication is defined as follows: let $\mathbf{q}$ and $\mathbf{p}$ be two quaternions having elements $\{a, b, c, d\}$ and $\{e, f, g, h\}$, respectively, then

$$\mathbf{q} \cdot \mathbf{p} = \begin{pmatrix} a & -b & -c & -d \\ b & a & -d & c \\ c & d & a & -b \\ d & -c & b & a \end{pmatrix} \begin{pmatrix} e \\ f \\ g \\ h \end{pmatrix} \tag{23}$$

The Euler and Cayley-Hamilton Theorems can be employed to derive multiple formulations for rotations relying on the fact that any rotation matrix $\mathbf{R}$ encodes a single axis of rotation which is the eigenvector $\mathbf{e} = \begin{pmatrix} e_1 \; e_2 \; e_3 \end{pmatrix}^T$ associated with the eigenvalue $+1$. Along with this, the following relation holds for a rotation ϕ about this axis:

$$\cos\phi = \frac{\text{trace}(\mathbf{R}) - 1}{2} \tag{24}$$

Through suitable derivation, we may define a rotation therefore by a four-parameter vector $\boldsymbol{\lambda}$:

$$\boldsymbol{\lambda} = \begin{pmatrix} \cos\phi \; e_1\sin\phi \; e_2\sin\phi \; e_3\sin\phi \end{pmatrix}^T \tag{25}$$

where the first element is the term involving the rotation and the last three define the *vector of the rotation matrix* which is sufficient for a single rotation but leaves the problem of propagating the transformation in time. A convenient relation between the elements of $\boldsymbol{\lambda}$ and quaternions exists, which allows us to take advantage of some felicitous properties of quaternions . Let $\mathbf{q}$ be the vector of the quaternion elements a, b, c, d as defined in (22). Then

$$a = \pm\sqrt{\frac{1+\lambda_1}{2}} \qquad = \cos(||\mathbf{e}||/2) \tag{26}$$

$$b = \frac{\lambda_2}{2a} \qquad =(e_1/||\mathbf{e}||)\sin(||\mathbf{e}||/2) \tag{27}$$

$$c = \frac{\lambda_3}{2a} \qquad =(e_2/||\mathbf{e}||)\sin(||\mathbf{e}||/2) \tag{28}$$

$$d = \frac{\lambda_4}{2a} \qquad =(e_3/||\mathbf{e}||)\sin(||\mathbf{e}||/2) \tag{29}$$

The parameters of the quaternion are properly called the *Euler-Rodrigues parameters* (Angeles, 2003) which define a unit quaternion. The rotation matrix in (20) in terms of Euler-Rodrigues parameters is

$$\mathbf{R}_b^n = \begin{pmatrix} (a^2+b^2-c^2-d^2) & 2(bc-ad) & 2(bd+ac) \\ 2(bc+ad) & (a^2-b^2+c^2-d^2) & 2(cd-ab) \\ 2(bd-ac) & 2(cd+ab) & (a^2-b^2-c^2+d^2) \end{pmatrix} \tag{30}$$

2.6.3 Time-derivative of the DCM

Let the vector $\boldsymbol{\omega}_{nb}^n$ be the rotation rates of the body axes about the navigation system axes expressed in the n-frame given by

$$\boldsymbol{\omega}_{nb}^n = \mathbf{R}_b^n \boldsymbol{\omega}_{nb}^b \tag{31}$$

and the resulting perpendicular linear velocity is given by

$$\begin{aligned} \dot{\mathbf{r}}^n &= \boldsymbol{\omega}_{nb}^n \times \mathbf{r}^b \\ &= \mathbf{R}_b^n \boldsymbol{\omega}_{nb}^b \times \mathbf{r}^b \\ &= \mathbf{R}_b^n \boldsymbol{\Omega}_{nb}^b \mathbf{r}^b \end{aligned} \tag{32}$$

where $\boldsymbol{\Omega}_{nb}^b$ is the skew-symmetric form of $\boldsymbol{\omega}_{nb}^b$, with elements

$$\boldsymbol{\Omega}_{nb}^b = \begin{pmatrix} 0 & -\omega_z & \omega_y \\ \omega_z & 0 & -\omega_x \\ -\omega_y & \omega_x & 0 \end{pmatrix} \tag{33}$$

We also have

$$\dot{\mathbf{r}}^n = \dot{\mathbf{R}}_b^n \mathbf{r}^b \tag{34}$$

Equating (32) and (34) yields the time derivative of the rotation matrix

$$\dot{\mathbf{R}}_b^n = \mathbf{R}_b^n \boldsymbol{\Omega}_{nb}^b \tag{35}$$

Over short periods of time for discrete measurements, the change in $\mathbf{R}_b^n$ can be computed using the small angle approximation of (20), where $\sin\varphi \approx \varphi$, $\sin\theta \approx \theta$ and $\sin\psi \approx \psi$, which is

$$\mathbf{R}_b^n \approx \begin{pmatrix} 1 & -\psi & \theta \\ \psi & 1 & -\varphi \\ -\theta & \varphi & 1 \end{pmatrix} \tag{36}$$

so that

$$\mathbf{R}_b^n(t+\delta t) = \mathbf{R}_b^n(t)\mathbf{R}_b^n(t, t+\delta t) \tag{37}$$

where $\mathbf{R}_b^n(t, t+\delta t)$ is the incremental rotation between the b and n-frames from time t to time $t+\delta t$. It is worth noting that under small incremental angles, the order of the rotations is not important.

3. Modeling

3.1 Linear motion

Using Newton's second law, the motion of a particle in the i-frame is given as

$$\ddot{\mathbf{r}}^i = \mathbf{f}^\mathbf{i} \tag{38}$$

where $\ddot{\mathbf{r}}^i$ is the second time derivative of position and $\mathbf{f}^\mathbf{i}$ is the specific force acting on the particle. Incorporating the accelerations due to gravitation, we have

$$\ddot{\mathbf{r}}^i = \mathbf{f}^i + \bar{\mathbf{g}}^i \tag{39}$$

Equation (39) represents a set of second-order differential equations which can be rewritten as a set of first-order equations:

$$\dot{\mathbf{r}}^i = \mathbf{v}^\mathbf{i} \tag{40}$$

$$\dot{\mathbf{v}}^i = \mathbf{f}^i + \bar{\mathbf{g}}^i \tag{41}$$

in which $\dot{\mathbf{r}}$, the first time derivative of position is equated with velocity $\mathbf{v}$. We now turn to the derivation of the model equations for navigating in the i-, e- and l-frames

3.2 State models for kinematic geodesy

3.2.1 The i-frame

Because a vehicle is oriented arbitrarily with respect to the i-frame as defined above, the measurements of specific force will not be in this frame, but rather in the body-frame.[4]. A rotation matrix $\mathbf{R}_b^i$ is used to resolve the forces in the i-frame:

$$\mathbf{f}^i = \mathbf{R}_b^i \mathbf{f}^b \tag{42}$$

In this notation, superscript of the measurement vector $\mathbf{f}$ and subscript of the rotation matrix cancel, yielding the representation of the vector in the desired frame. In navigation

[4] after the rigid transformation between the IMU and the vehicle has been applied

applications, the time derivative of $\mathbf{R_b^i}$ is a function of the angular velocity expressed by the vector $\boldsymbol{\omega}_{ib}^b$ between the two reference frames. Here, $\boldsymbol{\omega}_{ib}^b$ is the representation of the rotation rate expressed in the body frame whose skew-symmetric form is the matrix $\boldsymbol{\Omega}_{ib}^b$, giving

$$\dot{\mathbf{R}}_b^i = \mathbf{R}_b^i \boldsymbol{\Omega}_{ib}^b \tag{43}$$

which is, of course, a particular realization of (35), in which $n = i$. Assuming now that the gravity vector is computed by (11), we must apply a transformation from the l-frame to the i-frame to formulate (41) as

$$\dot{\mathbf{v}} = \mathbf{R}_b^i \mathbf{f}^b + \mathbf{R}_l^i \bar{\mathbf{g}}^l \tag{44}$$

The time derivatives of position, velocity and attitude are now represented as

$$\dot{\mathbf{x}}^i = \begin{pmatrix} \dot{\mathbf{r}}^i \\ \dot{\mathbf{v}}^i \\ \dot{\mathbf{R}}_b^i \end{pmatrix} = \begin{pmatrix} \mathbf{v}^i \\ \mathbf{R}_b^i \mathbf{f}^b + \mathbf{R}_l^i \bar{\mathbf{g}}^l \\ \mathbf{R}_b^i \boldsymbol{\Omega}_{ib}^b \end{pmatrix} \tag{45}$$

The solution to (45) is the navigation state of the vehicle: position, velocity and attitude in the inertial frame. The equations in (45) are known as the *mechanization equations* for the inertial navigation system.

3.2.2 The e-frame

To arrive at the state equations in the e-frame, we begin by considering the transformation of the position vector in the e-frame into the i-frame:

$$\mathbf{r}^i = \mathbf{R}_e^i \mathbf{r}^e \tag{46}$$

The first and second derivatives are then

$$\dot{\mathbf{r}}^i = \mathbf{R}_e^i \left(\dot{\mathbf{r}}^e + \boldsymbol{\Omega}_{ie}^e \mathbf{r}^e\right) \tag{47}$$

$$\ddot{\mathbf{r}}^i = \mathbf{R}_e^i \left(\ddot{\mathbf{r}}^e + 2\boldsymbol{\Omega}_{ie}^e \dot{\mathbf{r}}^e + \dot{\boldsymbol{\Omega}}_{ie}^e \mathbf{r}^e + \boldsymbol{\Omega}_{ie}^e \boldsymbol{\Omega}_{ie}^e \mathbf{r}^e\right) \tag{48}$$

As the rotation rate between the e and i frames is constant, we see that $\dot{\boldsymbol{\Omega}}_{ie}^e$ is zero leaving

$$\ddot{\mathbf{r}}^i = \mathbf{R}_e^i \left(\ddot{\mathbf{r}}^e + 2\boldsymbol{\Omega}_{ie}^e \dot{\mathbf{r}}^e + \boldsymbol{\Omega}_{ie}^e \boldsymbol{\Omega}_{ie}^e \mathbf{r}^e\right) \tag{49}$$

The second derivative of position in the e-frame is obtained using (10) and (39), yielding

$$\ddot{\mathbf{r}}^e = \mathbf{R}_b^e \mathbf{f}^b - 2\boldsymbol{\Omega}_{ie}^e \dot{\mathbf{r}}^e + \mathbf{g}^e \tag{50}$$

The second term in (50) is due to the Coriolis force, whereas the last term is the gravity vector represented in the e-frame, which can be found in (Schwarz & Wei, 1990).

Putting this together to form the state-variable equations, we have

$$\dot{\mathbf{x}}^e = \begin{pmatrix} \dot{\mathbf{r}}^e \\ \dot{\mathbf{v}}^e \\ \dot{\mathbf{R}}_b^e \end{pmatrix} = \begin{pmatrix} \mathbf{v}^e \\ \mathbf{R}_b^e \mathbf{f}^b - 2\boldsymbol{\Omega}_{ie}^e \mathbf{v}^e + \mathbf{g}^e \\ \mathbf{R}_b^e \left(\boldsymbol{\Omega}_{ie}^b - \boldsymbol{\Omega}_{ib}^b\right) \end{pmatrix} \tag{51}$$

3.2.3 The l-frame

Navigation states expressed in the inertial and earth-fixed frames do not lend themselves to easy intuitive interpretation near the surface of the earth. Here, the more familiar concepts of latitude and longitude along with roll, pitch and heading are preferable. We therefore must mechanize the system in the l-frame, which necessitates a reformulation of the state-variable equations. To begin with, we note

$$\mathbf{r}^l = (\phi\ \lambda\ h)^T \tag{52}$$

whose time rate is

$$\dot{\mathbf{r}}^l = (\dot{\phi}\ \dot{\lambda}\ \dot{h})^T \tag{53}$$

Rather than express velocity in terms of the geodetic coordinates, it is preferable to represent them in the *enu* system:

$$\mathbf{v}^l = \begin{pmatrix} v_e & v_n & v_u \end{pmatrix}^T \tag{54}$$

Now, the time derivative of position in ϕ, λ, h is related to $\mathbf{v}^l$ through

$$\dot{\mathbf{r}}^l = \mathbf{D}^{-1}\mathbf{v}^l \tag{55}$$

or

$$\begin{pmatrix} \dot{\phi} \\ \dot{\lambda} \\ \dot{h} \end{pmatrix} = \begin{pmatrix} 0 & \frac{1}{(M+h)} & 0 \\ \frac{1}{(N+h)\cos\phi} & 0 & 0 \\ 0 & 0 & 1 \end{pmatrix} \begin{pmatrix} v_e \\ v_n \\ v_u \end{pmatrix} \tag{56}$$

The first two non-zero elements of $\mathbf{D}^{-1}$ are clearly derivable from the derivatives of equations (6) and (7) with respect to time. Acceleration is now given by

$$\begin{aligned} \dot{\mathbf{v}}^l &= \mathbf{R}_i^l \ddot{\mathbf{r}}^i - (2\mathbf{\Omega}_{ie}^l + \mathbf{\Omega}_{el}^l)\mathbf{v}^l - \mathbf{\Omega}_{ie}^l \mathbf{\Omega}_{el}^l \mathbf{r}^l \\ &= \mathbf{R}_b^l \mathbf{f}^b - (2\mathbf{\Omega}_{ie}^l + \mathbf{\Omega}_{el}^l)\mathbf{v}^l + \mathbf{g}^l \end{aligned} \tag{57}$$

where $\mathbf{\Omega}_{ie}^l$ is the angular velocity of Earth's rotation expressed in the l-frame, $\mathbf{\Omega}_{el}^l$ is the angular velocity of the l-frame with respect to the e-frame expressed in the l-frame and $\mathbf{g}^l$ is as defined in (11). Finally, the transformation $\mathbf{R}_b^l$ is the solution to

$$\dot{\mathbf{R}}_b^l = \mathbf{R}_b^l \mathbf{\Omega}_{lb}^b = \mathbf{R}_b^l(\mathbf{\Omega}_{ib}^b - \mathbf{\Omega}_{il}^b) \tag{58}$$

The state-variable equations in the local-level reference frame are therefore

$$\dot{\mathbf{x}}^l = \begin{pmatrix} \dot{\mathbf{r}}^l \\ \dot{\mathbf{v}}^l \\ \dot{\mathbf{R}}_b^l \end{pmatrix} = \begin{pmatrix} \mathbf{D}^{-1}\mathbf{v}^l \\ \mathbf{R}_b^l \mathbf{f}^b - (2\mathbf{\Omega}_{ie}^l + \mathbf{\Omega}_{el}^l)\mathbf{v}^l + \mathbf{g}^l \\ \mathbf{R}_b^l(\mathbf{\Omega}_{ib}^b - \mathbf{\Omega}_{il}^b) \end{pmatrix} \tag{59}$$

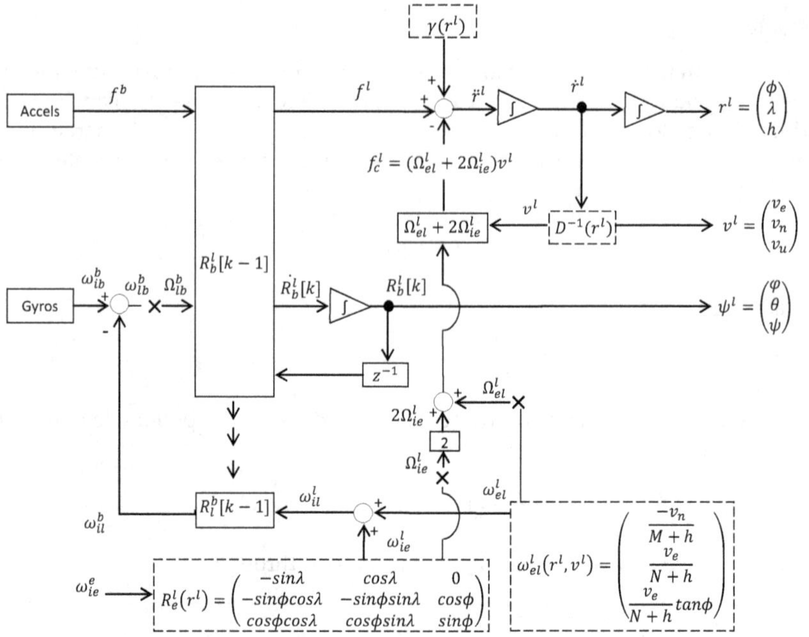

Fig. 7. Mechanization in the l-frame

3.3 Mechanization in the l-frame

Because of its wide applicability and intuitiveness, we shall focus on the mechanization of the state-variable equations described above in the local-level frame. To begin with, an initial position in geodetic coordinates ϕ, λ, h must be known, along with an initial velocity and transformation $\mathbf{R}_l^b$ in order for the integration of the measurements from the accelerometers and gyroscopes to give proper navigation parameters. We shall consider initial position and velocity to be given by GPS, for example, and will treat the problem of resolving initial attitude later. The block diagram in figure 7 shows the relationships among the components of the state-variable equations in the context of an algorithmic implementation.

Given an initial attitude, velocity and the earth's rotation rate $\boldsymbol{\omega}_e$, the rotation of the l-frame with respect to the e-frame and thence the rotation between the e-frame and the i-frame is computed and transformed into a representation of the rotation of the l-frame with respect to the i-frame *expressed in the b-frame* ($\boldsymbol{\omega}_{il}^b$). The quantities of this vector are subtracted from the body angular rate measurements to yield angular rates between the l-frame and the b-frame expressed in the b-frame ($\boldsymbol{\omega}_{lb}^b$). Given fast enough measurements relative to the dynamics of the vehicle, the small angle approximation can be used and $\mathbf{R}_l^b$ can be integrated over the time $\{t, t+\delta t\}$ to provide the next $\mathbf{R}_l^b$ which is used to transform the accelerometer measurements into the l frame. The normal gravity γ computed via Somigliana's formula (12) is added while the quantities arising from the Coriolis force are subtracted, yielding the acceleration in the l-frame. This, in turn, is integrated to provide velocity and again to yield position, which are fed back into the system to update the necessary parameters and propagate the navigation state forward in time.

3.4 Updating the transformation $\mathbf{R}_b^l$

The solution of (35) propagates the transformation matrix $\mathbf{R}_b^l$ in time. As both $\mathbf{R}_b^l$ and $\boldsymbol{\Omega}$ are time dependent, no closed form solution exists

(Kohler & Johnson, 2006). During a small time interval δt relative to the dynamics of the vehicle, however, we may assume a constant angular rate $\boldsymbol{\omega}$. The angular changes of the b-frame with respect to the l-frame are expressed as $\boldsymbol{\alpha} = \boldsymbol{\omega}\delta t$. The skew-symmetric form $\boldsymbol{\Omega}\delta t$ is now constant over a short time. This presents the discrete closed form solution

$$\mathbf{R}_{k+1} = \mathbf{R}_k e^{\boldsymbol{\Omega}\delta t} = \mathbf{R}_k \sum_{n=0}^{\infty} \frac{1}{n!}(\boldsymbol{\Omega}^n \delta t^n) = \mathbf{R}_k \sum_{n=0}^{\infty} \frac{1}{n!}\bar{\boldsymbol{\Omega}}^n \tag{60}$$

The powers of $\bar{\boldsymbol{\Omega}}$ are expressed as

$$\begin{aligned}
\bar{\boldsymbol{\Omega}}^3 &= -||\boldsymbol{\alpha}||^2\bar{\boldsymbol{\Omega}} \\
\bar{\boldsymbol{\Omega}}^4 &= -||\boldsymbol{\alpha}||^2\bar{\boldsymbol{\Omega}}^2 \\
\bar{\boldsymbol{\Omega}}^5 &= ||\boldsymbol{\alpha}||^4\bar{\boldsymbol{\Omega}} \\
\bar{\boldsymbol{\Omega}}^6 &= ||\boldsymbol{\alpha}||^4\bar{\boldsymbol{\Omega}}^2 \\
&\vdots
\end{aligned}$$

which allows us to collect the terms in (60) in sine and cosine components of the series expansion to obtain

$$\mathbf{R}_{k+1} = \mathbf{R}_k \left(\mathbf{I} + \frac{1}{||\boldsymbol{\alpha}||}\sin(||\boldsymbol{\alpha}||)\bar{\boldsymbol{\Omega}} + \frac{1-\cos(||\boldsymbol{\alpha}||)}{||\boldsymbol{\alpha}||^2}\bar{\boldsymbol{\Omega}}^2\right) \tag{61}$$

3.4.1 Quaternion update

The quaternion parametrization requires a four-element vector as we have seen. A discrete solution to the quaternion update follows similarly to that of the DCM. The difference here is that now the skew-symmetric representation of the rotation rate vector is 4×4. In block form, we have

$$\boldsymbol{\Omega}_q = \begin{pmatrix} -\boldsymbol{\Omega} & \boldsymbol{\omega} \\ -\boldsymbol{\omega}^T & 0 \end{pmatrix} \tag{62}$$

Letting $\bar{\boldsymbol{\Omega}}_q = \boldsymbol{\Omega}_q\delta t$, the discrete solution to (??) is

$$\begin{aligned}
\mathbf{q}_{k+1} &= \left(\sum_{n=0}^{\infty} \frac{1}{2^n n!}\bar{\boldsymbol{\Omega}}_q^n\right)\mathbf{q}_k \\
&= \mathbf{q}_k + \frac{1}{2}\left(2\left(\cos\frac{||\boldsymbol{\alpha}||}{2} - 1\right)\mathbf{I} + \frac{2}{||\boldsymbol{\alpha}||}\sin\frac{||\boldsymbol{\alpha}||}{2}\bar{\boldsymbol{\Omega}}_q\right)\mathbf{q}_k
\end{aligned} \tag{63}$$

Implementation in either the DCM or the quaternion parametrizations involves employing numerical integration techniques, which we shall not cover here.

3.5 Initialization

As stated above, the implementation of an INS requires the knowledge of initial position, velocity and attitude. Initial position and velocity can be provided through any appropriate means, but most commonly are retrieved through GPS measurements. The initial attitude, on the other hand, can be resolved using the raw measurements of the IMU and the known or computed gravity and earth rotation rate. """"'Initialization can be performed from a static position or during maneuvers, the latter being more complex and beyond the scope of this chapter.

3.5.1 Alignment of a static platform

Alignment refers to the process of determining the initial orientation of the INS body axes with respect to the navigation frame by rotating the system until expected measurements are observed in the transformed output. Specifically, with respect to the x and y axes, we define the process as *leveling*, while the heading (about the z axis) is termed *gyro-compassing*. First, we begin by noting that the measured specific forces in the body frame are related to gravity in the local-level frame through

$$\mathbf{g}^l = \mathbf{R}_b^l \mathbf{f}^b \tag{64}$$

Through the orthogonality of $\mathbf{R}_b^l$, we see that

$$\begin{pmatrix} R_{31} \\ R_{32} \\ R_{33} \end{pmatrix} = -\frac{1}{g} \begin{pmatrix} f_x \\ f_y \\ f_z \end{pmatrix} \tag{65}$$

which defines one basis vector of the transformation. The sensed rotation rates are similarly related to the earth rotation rate through

$$\boldsymbol{\omega}_{ie}^l = \mathbf{R}_b^l \boldsymbol{\omega}_{ie}^b \tag{66}$$

or

$$\begin{pmatrix} 0 \\ \omega_e \cos\phi \\ \omega_e \sin\phi \end{pmatrix} = \begin{pmatrix} R_{11} & R_{12} & R_{13} \\ R_{21} & R_{22} & R_{23} \\ R_{31} & R_{32} & R_{33} \end{pmatrix} \begin{pmatrix} \omega_x \\ \omega_y \\ \omega_z \end{pmatrix} \tag{67}$$

It is easy to show that

$$\begin{pmatrix} R_{21} \\ R_{22} \\ R_{23} \end{pmatrix} = \frac{1}{\omega_e \cos\phi} \begin{pmatrix} \omega_x \\ \omega_y \\ \omega_z \end{pmatrix} - \tan\phi \begin{pmatrix} R_{31} \\ R_{32} \\ R_{33} \end{pmatrix} \tag{68}$$

from which the second basis of the transformation is found. To complete the rotation matrix, we take advantage again of its orthogonality, arriving at

$$\begin{pmatrix} R_{11} \\ R_{12} \\ R_{13} \end{pmatrix} = \begin{pmatrix} R_{21} \\ R_{22} \\ R_{23} \end{pmatrix} \times \begin{pmatrix} R_{31} \\ R_{32} \\ R_{33} \end{pmatrix} \tag{69}$$

We have thereby resolved the planar tilt of the body frame with respect to the local-level frame as well as the rotation about the leveled z axis in the body frame that would bring about a zero-rate measurement along the transformed x axis of the IMU. In practice, sensor noise of vehicle disturbance would not allow for the exact solutions presented above, leading to an initial alignment error. One way to minimize the error is to collect stationary measurements over an extended period of time and compute the mean values or apply another type of low-pass filter. It is worth noting here that the estimate of this alignment step is considered coarse and can be further improved through a fine alignment process in which external aiding, in the form of position and/or velocity updates, are used.

3.6 Error dynamics

The state-variable equations described up to now for determining the navigation parameters of the vehicle represent non-linear dynamic system with the general form

$$\dot{\mathbf{x}}(t) = \mathbf{f}(t, \mathbf{x}(t), \mathbf{u}(t)) \tag{70}$$

where $\mathbf{x}$ are the physical parameters of the system and $\mathbf{u}$ are inputs to the system. The true values of $\mathbf{x}$ are generally not known, with only an approximation available. For example, in an INS, the approximation comes from the integration of sensor output over time. Let $\tilde{\mathbf{x}}$ represent the approximation, then the true parameters are

$$\tilde{\mathbf{x}} = \mathbf{x} + \delta\mathbf{x} \tag{71}$$

where $\delta\mathbf{x}$ are the error states. Replacing $\mathbf{x}$ with $\tilde{\mathbf{x}}$, we have

$$\dot{\tilde{\mathbf{x}}}(\mathbf{t}) = \mathbf{f}(t, \tilde{\mathbf{x}}(t), \mathbf{u}(t)) = \mathbf{f}(t, \mathbf{x}(t) + \delta\mathbf{x}(t), \mathbf{u}(t)) \tag{72}$$

Taylor series approximation to the error term yields

$$\delta\dot{\mathbf{x}}(t) = \frac{\partial \mathbf{f}}{\partial \mathbf{x}} \delta\mathbf{x}(t) \tag{73}$$

which is the linearized form of the state equations in terms of the errors, called the *error state equations*. It takes the familiar form

$$\delta\dot{\mathbf{x}}(t) = \mathbf{F}_x \delta\mathbf{x}(t) \tag{74}$$

Where $\mathbf{F}_x$ is the dynamics matrix. So far, we have a model for a purely deterministic system (one free of sensor errors). Taking the input as part of the (noisy) sensor output, the second set of differential equations is formed:

$$\delta\dot{\mathbf{u}} = \mathbf{F}_u \delta\mathbf{u}(t) + \mathbf{G}\mathbf{w}(t) \tag{75}$$

Where $\mathbf{F}_u$ is the dynamics matrix for the sensor errors, $\mathbf{w}(t)$ is a random Gaussian sequence with a shaping matrix $\mathbf{G}$. The general state-variable form of the error model is therefore

$$\underbrace{\begin{pmatrix} \delta\dot{\mathbf{x}} \\ \delta\dot{\mathbf{u}} \end{pmatrix}}_{\dot{\mathbf{x}}(t)} = \underbrace{\begin{pmatrix} \mathbf{F}_x & \mathbf{F}_{xu} \\ \mathbf{0} & \mathbf{F}_u \end{pmatrix}}_{\mathbf{F}} \underbrace{\begin{pmatrix} \delta\mathbf{x} \\ \delta\mathbf{u} \end{pmatrix}}_{\mathbf{x}(t)} + \underbrace{\begin{pmatrix} \mathbf{0} \\ \mathbf{G} \end{pmatrix}}_{\mathbf{G}} \mathbf{w} \tag{76}$$

The terms in $\mathbf{F}_{xu}$ account for the dependence of the navigational errors upon the sensor errors and the full state vector includes the elements of $\mathbf{u}$. In an INS mechanized in the l-frame, the error state vector is explicitly written

$$\mathbf{x}(t) = \left(\delta\phi \; \delta\lambda \; \delta h \; \delta v_e \; \delta v_n \; \delta v_u \; \epsilon_e \; \epsilon_n \; \epsilon_u \; d_x \; d_y \; d_z \; b_x \; b_y \; b_z \right)^T \tag{77}$$

where the first three elements are position errors, the next three are velocity errors and after which come alignment errors, gyro drifts and accelerometer biases. We shall now derive the equations governing each.

3.6.1 Position errors

Recall that in section 3.2.3, we preferentially expressed velocities in the l-frame in terms of v_e, v_n, v_u rather than directly as functions of ϕ, λ, h, using

$$\mathbf{v}^l = \mathbf{D}\dot{\mathbf{r}}^l \tag{78}$$

where

$$\mathbf{D} = \begin{pmatrix} 0 & (N+h)\cos\phi & 0 \\ (M+h) & 0 & 0 \\ 0 & 0 & 1 \end{pmatrix} \tag{79}$$

which, after linearization, yields the error equation

$$\delta\mathbf{v}^l = \mathbf{D}\delta\dot{\mathbf{r}}^l + \delta\mathbf{D}\dot{\mathbf{r}}^l \tag{80}$$

Recognizing that the first term in (80) is simply $\mathbf{D}\delta\mathbf{v}^l$, we see that it is the second term that contains the position errors. We can rewrite (80) then as

$$\delta\mathbf{v}^l = \mathbf{D}\delta\dot{\mathbf{r}}^l + \mathbf{D}_r\delta\mathbf{r}^l \tag{81}$$

where $\mathbf{D}_r$ is a coefficient matrix. The position error equation is then

$$\delta\dot{\mathbf{r}}^l = \mathbf{D}^{-1}\delta\mathbf{v}^l - \mathbf{D}^{-1}\mathbf{D}_r\delta\mathbf{r}^l \tag{82}$$

3.6.2 Velocity errors

The approximate form of (57) is

$$\begin{aligned} \dot{\tilde{\mathbf{v}}}^l &= \tilde{\mathbf{R}}_b^l\tilde{\mathbf{f}}^b - (2\tilde{\mathbf{\Omega}}_{ie}^l + \tilde{\mathbf{\Omega}}_{el}^l)\tilde{\mathbf{v}}^l + \tilde{\gamma}^l \\ &= (\mathbf{I} + \mathbf{E}^l)\mathbf{R}_b^l(\mathbf{f}^b + \delta\mathbf{f}^b) - (2(\mathbf{\Omega}_{ie}^l + \delta\mathbf{\Omega}_{ie}^l) + \mathbf{\Omega}_{el}^l + \delta\mathbf{\Omega}_{el}^l)(\mathbf{v}^l + \delta\mathbf{v}^l) + (\gamma^l + \delta\gamma^l) \end{aligned} \tag{83}$$

where $\mathbf{E}^l$ is the skew-symmetric form of the alignment error. After subtracting the true acceleration and ignoring second-order terms, we have

$$\begin{aligned} \delta\dot{\mathbf{v}}^l &= \mathbf{E}^l\mathbf{R}_b^l\mathbf{f}^b - (2\mathbf{\Omega}_{ie}^l + \mathbf{\Omega}_{el}^l)\delta\mathbf{v}^l - (2\delta\mathbf{\Omega}_{ie}^l + \delta\mathbf{\Omega}_{el}^l)\mathbf{v}^l + \delta\gamma^l + \mathbf{R}_b^l\delta\mathbf{f}^b - \delta\mathbf{g}^l \\ &= -\mathbf{F}^l\epsilon^l - (2\mathbf{\Omega}_{ie}^l + \mathbf{\Omega}_{el}^l)\delta\mathbf{v}^l + \mathbf{V}^l(2\delta\omega_{ie}^l + \delta\omega_{el}^l) + \delta\gamma^l + \mathbf{R}_b^l\mathbf{b} \end{aligned} \tag{84}$$

where $\mathbf{F}^l$ is the skew-symmetric matrix representation of $\mathbf{f}^l$, ϵ^l is the misalignment vector, $\mathbf{V}^l$ is the skew-symmetric form of $\mathbf{v}^l$ and $\mathbf{b}$ is the vector of accelerometer biases, in which the gravity disturbance vector $\delta\mathbf{g}^l$ is also included. The terms $\delta\omega_{el}^l$ and $\delta\omega_{ie}^l$ are the errors in the transport rate and the Earth rotation rate, respectively, both of which are dependent on errors in position and velocity. Finally, $\delta\gamma^l$ is the error in the calculation of normal gravity, which is dependent on the position error and any errors in the model.

3.6.3 Alignment errors

The alignment errors $\boldsymbol{\epsilon}^l$ represent misalignment between the b and l frames expressed in the l-frame. The vector $\boldsymbol{\epsilon}^l$ can be expressed in skew-symmetric form as $\mathbf{E}^l$, so that the approximate transformation between frames is

$$\tilde{\mathbf{R}}_b^l = (\mathbf{I} + \mathbf{E}^l)\mathbf{R}_b^l \tag{85}$$

The differential equations for the alignment errors are

$$\dot{\boldsymbol{\epsilon}}^l = -\boldsymbol{\Omega}_{il}^l \boldsymbol{\epsilon}^l - \delta\boldsymbol{\omega}_{il}^l - \mathbf{R}_b^l \mathbf{d} \tag{86}$$

where $\boldsymbol{\Omega}_{il}^l$ is the skew-symmetric form of the angular rates $\boldsymbol{\omega}_{il}^l$ with corresponding errors $\delta\boldsymbol{\omega}_{il}^l$. Here, $\mathbf{d}$ is the vector of gyro drift biases.

3.6.4 Gyroscope drifts and accelerometer biases

Gyroscopes and accelerometers exhibit noise behavior that is characterizeable at different time scales such that one can generally separate errors that are long-term stable and those that behave stochastically during the period of interest. Errors of the former type are characterized in a laboratory setting, prior to field deployment and their effects can generally be removed from the measurements, leaving residual errors that are modeled stochastically.

The noise in gyro and accelerometer measurements exhibit varying degrees of temporal correlation, depending on the quality of the devices. The underlying random processes are therefore conveniently modeled as first-order Gauss-Markov processes. Their equations are

$$\dot{\mathbf{d}} = -\boldsymbol{\alpha}\mathbf{d} + \mathbf{w}_d \tag{87}$$

$$\dot{\mathbf{b}} = -\boldsymbol{\beta}\mathbf{b} + \mathbf{w}_b \tag{88}$$

where $\boldsymbol{\alpha}$ and $\boldsymbol{\beta}$ are diagonal matrices whose non-zero elements are the reciprocals of the correlation time constants and $\mathbf{w}_d$ and $\mathbf{w}_b$ are white noise sequences.

3.6.5 Error state equations in the l frame

Combining the derivations for the error equations of position, velocity, alignment, gyro drift and accelerometer bias gives the state-variable equations for the navigation errors:

$$\dot{\mathbf{x}}^l(t) = \begin{pmatrix} \delta\dot{\mathbf{r}}^l \\ \delta\dot{\mathbf{v}}^l \\ \dot{\boldsymbol{\epsilon}}^l \\ \dot{\mathbf{d}} \\ \dot{\mathbf{b}} \end{pmatrix} = \begin{pmatrix} \mathbf{D}^{-1}\delta\mathbf{v}^l - \mathbf{D}^{-1}\mathbf{D}_r\delta\mathbf{r}^l \\ -\mathbf{F}^l\boldsymbol{\epsilon}^l - (2\boldsymbol{\Omega}_{ie}^l + \boldsymbol{\Omega}_{el}^l)\delta\mathbf{v}^l + \mathbf{V}^l(2\boldsymbol{\omega}_{ie}^l + \delta\boldsymbol{\omega}_{el}^l) + \delta\boldsymbol{\gamma}^l + \mathbf{R}_b^l\mathbf{b} \\ -\boldsymbol{\Omega}_{il}^l\boldsymbol{\epsilon}^l - \delta\boldsymbol{\omega}_{il}^l - \mathbf{R}_b^l\mathbf{d} \\ -\boldsymbol{\alpha}\mathbf{d} + \mathbf{w}_d \\ -\boldsymbol{\beta}\mathbf{b} + \mathbf{w}_b \end{pmatrix} \tag{89}$$

To derive the elements of the dynamics matrix $\mathbf{F}$, we need to specify all the matrix elements in (89).

3.6.6 Matrix formulation of position errors

Assuming M and N to be constant over small distances

$$\begin{aligned}\delta\mathbf{v}^l &= \mathbf{D}\delta\dot{\mathbf{r}}^l + \mathbf{D}_r\delta\mathbf{r}^l \\ &= \begin{pmatrix} 0 & (N+h)\cos\phi & 0 \\ M+h & 0 & 0 \\ 0 & 0 & 1 \end{pmatrix}\begin{pmatrix}\delta\dot{\phi} \\ \delta\dot{\lambda} \\ \delta\dot{h}\end{pmatrix} + \begin{pmatrix} -\dot{\lambda}(N+h)\sin\phi & 0 & \dot{\lambda}\cos\phi \\ 0 & 0 & \dot{\phi} \\ 0 & 0 & 0 \end{pmatrix}\begin{pmatrix}\delta\phi \\ \delta\lambda \\ \delta h\end{pmatrix}\end{aligned} \tag{90}$$

With $\mathbf{D}_r$ in hand,

$$\delta\dot{\mathbf{r}}^l = \underbrace{\begin{pmatrix} 0 & \frac{1}{M+h} & 0 \\ \frac{1}{(N+h)}\cos\phi & 0 & 0 \\ 0 & 0 & 1 \end{pmatrix}}_{\mathbf{F}_{12}}\begin{pmatrix}\delta v_e \\ \delta v_n \\ \delta v_u\end{pmatrix} - \underbrace{\begin{pmatrix} 0 & 0 & \frac{-\dot{\phi}}{M+h} \\ \dot{\lambda}\tan\phi & 0 & \frac{-\dot{\lambda}}{N+h} \\ 0 & 0 & 0 \end{pmatrix}}_{\mathbf{F}_{11}}\begin{pmatrix}\delta\phi \\ \delta\lambda \\ \delta h\end{pmatrix} \tag{91}$$

where $\mathbf{F}_{11}$ and $\mathbf{F}_{12}$ are the first two 3×3 sub-matrices of $\mathbf{F}$.

3.6.7 Matrix formulation of velocity errors

Next, we turn to the first term in the second equation of (89):

$$-\mathbf{F}^l\boldsymbol{\epsilon}^l = \underbrace{\begin{pmatrix} 0 & f_u & -f_n \\ -f_u & 0 & f_e \\ f_n & -f_e & 0 \end{pmatrix}}_{\mathbf{F}_{23}}\begin{pmatrix}\epsilon_e \\ \epsilon_n \\ \epsilon_u\end{pmatrix} \tag{92}$$

The second term is

$$-(2\boldsymbol{\Omega}^l_{ie} + \boldsymbol{\Omega}^l_{el})\delta\mathbf{v}^l = \begin{pmatrix} 0 & (2\omega_e+\dot{\lambda})\sin\phi & -(2\omega_e+\dot{\lambda})\cos\phi \\ -(2\omega_e+\dot{\lambda})\sin\phi & 0 & -\dot{\phi} \\ (2\omega_e+\dot{\lambda})\cos\phi & \dot{\phi} & 0 \end{pmatrix}\begin{pmatrix}\delta v_e \\ \delta v_n \\ \delta v_u\end{pmatrix} \tag{93}$$

The third term is as follows:

$$\begin{aligned}2\delta\boldsymbol{\omega}^l_{ie} + \delta\boldsymbol{\omega}^l_{el} &= \begin{pmatrix} -\delta\phi \\ -(2\omega_e+\dot{\lambda})\sin\phi\delta\phi + \cos\phi\delta\dot{\lambda} \\ (2\omega_e+\dot{\lambda})\cos\phi\delta\phi + \sin\phi\delta\dot{\lambda} \end{pmatrix} \\ &= \begin{pmatrix} 0 & 0 & \frac{\dot{\phi}}{M+h} \\ -2\omega_e\sin\phi & 0 & \frac{-\dot{\lambda}\cos\phi}{N+h} \\ 2\omega_e\cos\phi + \frac{\dot{\lambda}}{\cos\phi} & 0 & \frac{-\dot{\lambda}\sin\phi}{N+h} \end{pmatrix}\begin{pmatrix}\delta\phi \\ \delta\lambda \\ \delta h\end{pmatrix} + \begin{pmatrix} 0 & \frac{-1}{M+h} & 0 \\ \frac{1}{N+h} & 0 & 0 \\ \frac{\tan\phi}{N+h} & 0 & 0 \end{pmatrix}\begin{pmatrix}\delta v_e \\ \delta v_n \\ \delta v_u\end{pmatrix}\end{aligned} \tag{94}$$

Multiplying by $\mathbf{V}^l$ yields

$$\mathbf{V}^l(2\delta\boldsymbol{\omega}^l_{ie} + \delta\boldsymbol{\omega}^l_{el}) =$$

$$\begin{pmatrix} 2\omega_e(v_u\sin\phi + v_n\cos\phi) + v_n\frac{\dot{\lambda}}{\cos\phi} & 0 & 0 \\ -2\omega_e v_e\cos\phi - \frac{v_e\dot{\lambda}}{\cos\phi} & 0 & 0 \\ -2\omega_e v_e\sin\phi & 0 & 0 \end{pmatrix}\begin{pmatrix}\delta\phi \\ \delta\lambda \\ \delta h\end{pmatrix} + \begin{pmatrix} \frac{-v_u+v_n\tan\phi}{N+h} & 0 & 0 \\ \frac{-v_e\tan\phi}{N+h} & \frac{-v_u}{M+h} & 0 \\ \frac{v_e}{N+h} & \frac{v_n}{M+h} & 0 \end{pmatrix}\begin{pmatrix}\delta v_e \\ \delta v_n \\ \delta v_u\end{pmatrix} \tag{95}$$

The fourth term is

$$\delta\gamma^l = \begin{pmatrix} 0 \\ 0 \\ -\frac{\partial\gamma}{\partial h}\delta h \end{pmatrix} = \begin{pmatrix} 0 \\ 0 \\ \frac{2\gamma}{R_e \delta h} \end{pmatrix} \tag{96}$$

where R_e is the mean Earth radius. The last term is the transformed accelerometer biases. Combining all the terms, we have

$$\begin{aligned}\delta\dot{\mathbf{v}}^l = & \underbrace{\begin{pmatrix} 2\omega_e(v_u \sin\phi + v_n \cos\phi) + \frac{v_n\dot{\lambda}}{\cos\phi} & 0 & 0 \\ -2\omega_e v_e \cos\phi - \frac{v_e\dot{\lambda}}{\cos\phi} & 0 & 0 \\ -2\omega_e v_e \sin\phi & 0 & \frac{2\gamma}{R_e} \end{pmatrix}}_{\mathbf{F}_{21}} \delta\mathbf{r}^l + \\ & \underbrace{\begin{pmatrix} \frac{-\dot{h}+\dot{\phi}\tan\phi(M+h)}{N+h} & (2\omega_e + \dot{\lambda})\sin\phi & -(2\omega_e+\dot{\lambda})\cos\phi \\ -2(\omega_e+\dot{\lambda})\sin\phi & \frac{-\dot{h}}{M+h} & -\dot{\phi} \\ 2(\omega_e+\dot{\lambda})\cos\phi & 2\dot{\phi} & 0 \end{pmatrix}}_{\mathbf{F}_{22}} \delta\mathbf{v}^l + \\ & \underbrace{\begin{pmatrix} 0 & f_u & -f_n \\ -f_u & 0 & f_e \\ f_n & -f_e & 0 \end{pmatrix}}_{\mathbf{F}_{23}} \epsilon^l + \underbrace{\begin{pmatrix} R_{11} & R_{12} & R_{13} \\ R_{21} & R_{22} & R_{23} \\ R_{31} & R_{32} & R_{33} \end{pmatrix}}_{\mathbf{F}_{25}} \mathbf{b} \end{aligned} \tag{97}$$

3.6.8 Matrix formulation of alignment errors

The third equation in (89) can be derived similarly as

$$\begin{aligned}\dot{\epsilon}^l = & \underbrace{\begin{pmatrix} 0 & (\omega_e+\dot{\lambda})\sin\phi & -(\omega_e+\dot{\lambda})\cos\phi \\ -(\omega_e+\dot{\lambda})sin\phi & 0 & -\dot{\phi} \\ (\omega_e+\dot{\lambda})\cos\phi & \dot{\phi} & 0 \end{pmatrix}}_{\mathbf{F}_{33}} \epsilon^l + \\ & \underbrace{\begin{pmatrix} 0 & 0 & \frac{\dot{\phi}}{M+h} \\ -\omega_e\sin\phi & 0 & \frac{-\dot{\lambda}\cos\phi}{N+h} \\ \omega_e\cos\phi + \frac{\dot{\lambda}}{\cos\phi} & 0 & \frac{-\dot{\lambda}\sin\phi}{N+h} \end{pmatrix}}_{\mathbf{F}_{31}} \delta\mathbf{r}^l + \\ & \underbrace{\begin{pmatrix} 0 & \frac{-1}{M+h} & 0 \\ \frac{1}{N+h} & 0 & 0 \\ \frac{\tan\phi}{N+h} & 0 & 0 \end{pmatrix}}_{\mathbf{F}_{32}} \delta\mathbf{v}^l + \underbrace{\begin{pmatrix} R_{11} & R_{12} & R_{13} \\ R_{21} & R_{22} & R_{23} \\ R_{31} & R_{32} & R_{33} \end{pmatrix}}_{\mathbf{F}_{34}} \mathbf{d} \end{aligned} \tag{98}$$

3.6.9 Matrix formulation of sensor errors

The matrices associated with the gyro drift and accelerometer bias equations in (89) are diagonal, given as

$$\mathbf{F}_{44} = \begin{pmatrix} -\alpha_x & 0 & 0 \\ 0 & -\alpha_y & 0 \\ 0 & 0 & -\alpha_z \end{pmatrix} \tag{99}$$

$$\mathbf{F}_{55} = \begin{pmatrix} -\beta_x & 0 & 0 \\ 0 & -\beta_y & 0 \\ 0 & 0 & -\beta_z \end{pmatrix} \tag{100}$$

where the α and β terms are the reciprocals of the time constants associated with the first-order Gauss-Markov model of each sensor.

Finally, we can define the error dynamics matrix $\mathbf{F}$ in terms of the sub-matrices derived above as

$$\mathbf{F} = \begin{pmatrix} \mathbf{F}_{11} & \mathbf{F}_{12} & \mathbf{0} & \mathbf{0} & \mathbf{0} \\ \mathbf{F}_{21} & \mathbf{F}_{22} & \mathbf{F}_{23} & \mathbf{0} & \mathbf{F}_{25} \\ \mathbf{F}_{32} & \mathbf{F}_{32} & \mathbf{F}_{33} & \mathbf{F}_{34} & \mathbf{0} \\ \mathbf{0} & \mathbf{0} & \mathbf{0} & \mathbf{F}_{44} & \mathbf{0} \\ \mathbf{0} & \mathbf{0} & \mathbf{0} & \mathbf{0} & \mathbf{F}_{55} \end{pmatrix} \tag{101}$$

3.7 Error analysis and Schüler oscillation

In figure 7, it is shown that the rotation rate of the l-frame with respect to the i-frame expressed as a vector in the b-frame is subtracted from the raw gyroscope measurements when the system is to be mechanized in the l frame. The relative rotations between the frames, or the transport rate, itself is a function of the computed velocity and misalignment between the l and e frames. If there is an error in the computed transformation $\mathbf{R}_l^e$ or in the initial values in $\mathbf{R}_l^b$, the computation of $\mathbf{R}_b^l$ will be in error. In the simple case that the vehicle is actually perfectly level and either stationary or at a constant velocity, but the computed value of $\mathbf{R}_b^l$ indicates that it is not level, a component of the gravity vector will be resolved in the horizontal axes of the system. This component is integrated and provides an erroneous velocity value, which is fed back to compute ω_{il}^l, which, in turn is transformed through $\mathbf{R}_l^b$ and is subtracted from the angular rate measurements. Finally, a second integration occurs using the "corrected" measurements to update $\mathbf{R}_b^l$ and the process repeats.

The dynamics of the system described above are described by the characteristic equation in the Laplace domain as

$$s^2 + \frac{g}{R_e} = 0 \tag{102}$$

where g is gravity and R_e is the mean radius of the earth. This represents a simple oscillation whose natural frequency is

$$\omega_0 = \sqrt{\frac{g}{R_e}} \tag{103}$$

and is called the Schüler oscillation, after Maximilian Schüler who showed that the bob of a hypothetical pendulum whose string was the length of the Earth's radius would not be displaced under sudden motions of its support. The period of such a pendulum (and of the Schüler oscillation) is 84.4 minutes. This implies that positional errors caused by either accelerometer bias or initial velocity errors are bounded over this period. On the other hand, positional errors due to misalignment or gyro drift are not bounded.

Characterization of INS errors in each channel (East, North, Up) can be performed analytically in the case of a level platform traveling at a constant velocity and height where there is no coupling between them. For example, under the conditions stated above, a derivation of error propagation for the North channel proceeds from formulating an error dynamics matrix composed only of terms affecting the position, velocity and error states relating to it and deriving the state transition matrix $\mathbf{\Phi}$ by

$$\mathbf{\Phi}(t) = \mathcal{L}^{-1}\{(s\mathbf{I} - \mathbf{F})^{-1}\} \tag{104}$$

where $\mathcal{L}^{-1}$ denotes the inverse Laplace transform. The effect of a particular error source upon the error state under investigation is simply the term in $\mathbf{\Phi}(t)$ whose row index corresponds to the index of the state whose column index corresponds to the index of the error source. For example, the effect of a constant velocity error δv upon the North position is

$$\delta r_n(t) = \delta v \frac{\sin \omega_0 t}{\omega_0} \tag{105}$$

where ω_0 is the Schüler frequency. Because the errors in position due to gyro drift and misalignment are unbounded, as previously mentioned, the largest single quantity of merit in the sensor specifications of an IMU is in the gyro drift rate. We finish by noting that for more general trajectories, characterization of error propagation is best done through simulation.

4. Sensors

IMU measurements are made from two triads of orthogonally-mounted accelerometers and gyros; one sensor for each degree of freedom in three-dimensional space. Accelerometers measure *specific force* along a sensitive axis. Gyroscopes measure either rotations or rotation *rates* along a sensitive axis. Although the current state of the art in sensor design makes use of the principles of lasers and quantum mechanics, Newtonian mechanics gives us the tools to design both accelerometers and gyroscopes. Presented below is a brief discussion of the basis of such classical designs.

4.1 Accelerometers

Recalling Newton's second law:

$$\mathbf{F}^i = m\ddot{\mathbf{r}}^i \tag{106}$$

where i represents the inertial frame, expresses the fact that force is proportional to the acceleration of a constant proof mass. Conversely, the force needed to keep this mass from accelerating is a measure of linear acceleration, a principle employed in most accelerometers. It can be seen as a realization of the law of conservation of linear momentum:

$$\mathbf{F}^i = \dot{\mathbf{p}}^i = m\ddot{\mathbf{r}}^i = 0 \tag{107}$$

where $\mathbf{p}$ is the momentum of the proof mass, i.e. the rate of change of the momentum is equal to the applied force. The external forces acting on the system are balanced by internal forces, so the motion of the proof mass remains constant in an inertial frame. In theory there is a problem realizing such a sensor on Earth because the planet is undergoing constant acceleration in its orbit around the Sun and so forth. This makes defining zero acceleration impossible in an inertial frame, but we can simply treat any signal arising from these conditions as a constant instrument bias and remove it from the measurements. From here on, we will only consider this situation.

To see how (107) can be realized in a measurement device, consider the classical spring-mass-damper system shown in figure 8. A mass m is constrained to move along the

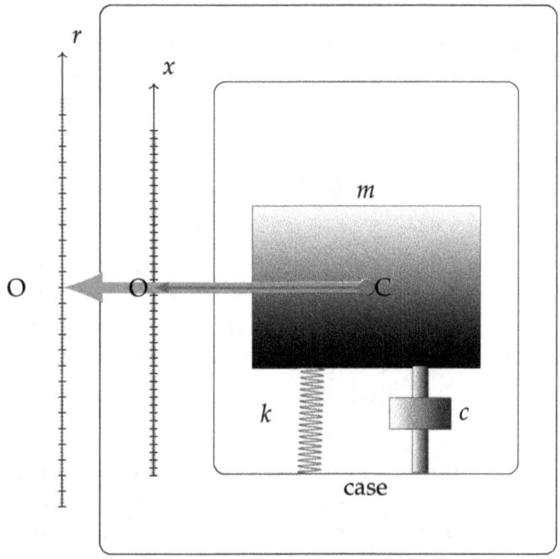

Fig. 8. a spring-mass-damper system

x axis of the device (the sensitive axis). It is restrained by a spring and its motion is damped by a damping device. Finally, there is a scale and a housing for the assembly. Point C is the center of mass of the sensitive element and point O indicates the equilibrium position when the device is not subjected to any external force along the sensitive axis. The output of the device is measured along the r scale, which is made proportional to the internal signal along x. The spring provides a restoring force proportional to the displacement of the proof mass by Hooke's Law:

$$F = -kx \tag{108}$$

where k is the spring constant. The damper is present to minimize oscillations in response to sudden changes in applied force and can be made of a viscous fluid-filled piston or the like. The force produced by the damper is proportional to the velocity of the proof mass or $-c\dot{x}$, where c is the damping or viscosity constant. If we assume that the sensor is located on the Earth with the x axis facing opposite the direction of the pull of gravity, Newton's second law gives the second-order differential equation

$$m\ddot{r}_C + c\dot{x}_C + kx_C - mg = 0 \tag{109}$$

where g is present to account for the reaction force of the case against the surface of the Earth. Letting

$$r_C = x_C + r_0 \tag{110}$$

we have

$$\ddot{x}_C + \frac{c}{m}\dot{x}_C + \frac{k}{m}x_C = g - \ddot{r}_0 \tag{111}$$

This system will therefore have an output of $g - \ddot{r}_0 = f$, which is the **specific force**. This is the observable obtained from accelerometers near the surface of Earth. In this case, without extra applied force, the output is simply g.

It is not possible to separate the effects of inertia and gravity in a non-inertial frame, a consequence of Einstein's equivalence principle. In other words, forces applied to an accelerometer through accelerations of the vehicle are indistinguishable from the acceleration caused by the gravity field of the planet. Without knowledge of the vehicle's acceleration at a particular time, it is not possible to measure the local gravitational vector and vice versa. The forces acting on the vehicle other than gravity include those induced by Earth's rotation, so we must be careful in how we eliminate instrument biases depending upon which reference frame we are to work in.

Equation (109) is an open-loop mechanization of the mass-spring-damper system, where the displacement is directly measured. Modern high-accuracy designs are by contrast closed-loop systems, where the mass is kept at the null position by a coil in a magnetic field. The force required to keep the mass stationary under various accelerations is then the quantity that is measured. Several other realizations of accelerometers are possible, but most are still modeled by similar differential equations.

Finally, we note that though the observable we shall deal with is specific force, the actual output of the sensor is change in velocity Δv. This is a consequence of the internal mechanisms of modern accelerometers, where several measurements are integrated over a short period of time (usually a few milliseconds) to smooth out measurement noise. The general form of the measurement model of specific force from an accelerometer triad is given by the observation equation

$$\ell_a = \mathbf{f} + \mathbf{b} + (\mathbf{S}_1 + \mathbf{S}_2)\mathbf{f} + \mathbf{N}\mathbf{f} + \gamma + \delta\mathbf{g} + \epsilon_f \tag{112}$$

where

ℓ_a is the measurement

$\mathbf{f}$ is the specific force

$\mathbf{b}$ is the accelerometer bias

$\mathbf{S}_1$ and $\mathbf{S}_2$ represent the linear and non-linear matrix of scale factor errors, respectively

N is a matrix representing the non-orthogonality of the sensor axes

γ is the vector of normal gravity

$\delta\mathbf{g}$ is the anomalous gravity vector

ϵ_f is noise

4.2 Gyroscopes

Gyroscopes measure angular velocity with respect to an inertial reference frame. A schematic of a simple two-axis gyroscope is shown in figure 9. In this device, a spinning disc is mounted within a set of gimbals which allow it to pivot in response to an applied torque, a behavior known as **precession**.

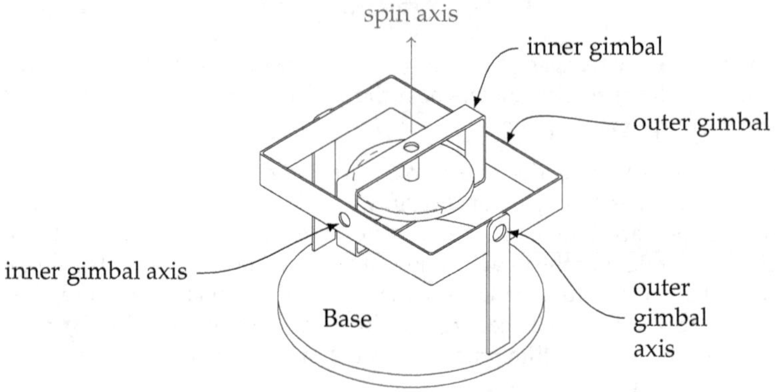

Fig. 9. a two-axis rigid rotor gyroscope

We can analyze the behavior of this system beginning with Newton's second law in terms of momentum again:

$$\mathbf{F}^i = \dot{\mathbf{p}}^i \tag{113}$$

The cross product with the vector **r** gives the moment of this force about the origin, or

$$\mathbf{r}^i \times \mathbf{F}^i = \mathbf{r}^i \times \dot{\mathbf{p}}^i \tag{114}$$

Now, we observe that the angular momentum, **L** of the spinning disc is

$$\mathbf{L} = \mathbf{r}^i \times \mathbf{p}^i \tag{115}$$

the time derivative of which is

$$\dot{\mathbf{L}} = \dot{\mathbf{r}}^i \times \mathbf{p}^i + \mathbf{r}^i \times \dot{\mathbf{p}}^i \tag{116}$$

Now, because $\dot{\mathbf{r}}^i$ and $\mathbf{p}^i$ are parallel, the first cross product on the right hand side of (116) is zero, thus

$$\mathbf{r}^i \times \mathbf{F}^i = \dot{\mathbf{L}} \tag{117}$$

For a particle moving in a central field (*i.e.* any point we chose on the disc), **F** and **r** are parallel and thus **L** is constant. This means that the direction of the spin axis of the rotating disc is fixed in inertial space. In a two axis gyroscope any rotation ω_t about t (the input axis) in figure 10 would give rise to a rotation ω_p about p (the output axis). This phenomenon is known as precession. Measuring the torque about p leads us to the angular velocity about t, which is the observable under consideration. As with accelerometers, the actual

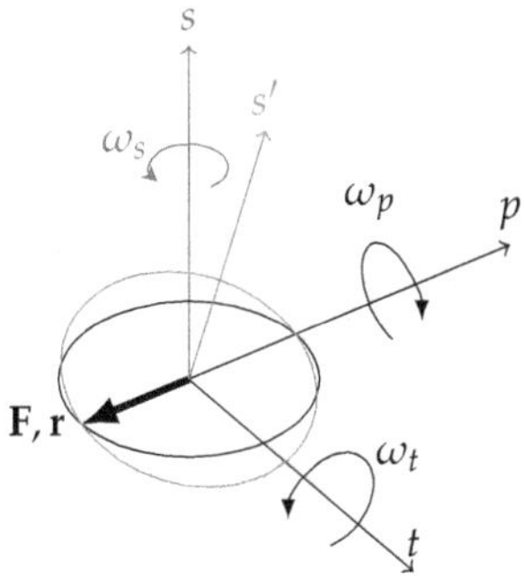

Fig. 10. gyroscopic precession

physical implementation of gyroscopes has taken on many forms, depending on purpose and performance considerations. In the example given above, measurements of gimbal rotation (in an open-loop system) are angular measurements. In a closed-loop system, motors are used to keep the gimbals from moving and the required torque to do so is measured. These measurements are therefore of the angular *rates* of the system. Sensing angular velocity in modern strap-down navigation systems is actually accomplished through exploiting the Sagnac effect rather than the mechanical properties of rotating masses. In this case, the interference patterns generated by light traveling along opposing closed paths is used as a measure of the angular rotation of the system. In any case, the measurements obtained from a gyroscope triad can modeled by the observation equation

$$\ell_\omega = \omega + d + \mathbf{S}\omega + \mathbf{N}\omega + \epsilon_\omega \tag{118}$$

where

ℓ_ω is the measurement

ω is the angular velocity

d is the gyroscope bias

S is a matrix representing the gyroscope scale factor

N is a matrix representing the non-orthogonality of the axes

ϵ_ω is noise

The noise terms of both accelerometers and gyroscopes can be further decomposed as

$$\epsilon = \epsilon_w + \epsilon_c + \epsilon_r + \epsilon_q + \epsilon_d \tag{119}$$

the five terms representing white, correlated, random walk, quantization and dither noise, respectively. Some IMU errors associated with scale factor and non-orthogonality are characterized in the factory, while others, including bias and noise are removed by the estimation process.

5. Estimation

5.1 Bayesian estimation

We now turn to the treatment of the stochastic aspects of INS design. In general, the dynamic system derived in the previous sections which describes the navigation and error states evolves in discrete time according to

$$\mathbf{x}_k = \mathbf{f}_{k-1}(\mathbf{x}_{k-1}, \mathbf{w}_{k-1}) \tag{120}$$

where $\mathbf{f}_{k-1}$ is some (possibly nonlinear) function of the previous state and its process noise $\mathbf{w}_{k-1}$, which accounts for errors in the model or disturbances to it. Also, generally speaking, we have no direct knowledge of the states themselves, but can only access them through measurements $\mathbf{z}$ which are related through

$$\mathbf{z}_k = \mathbf{h}_k(\mathbf{x}_k, \mathbf{v}_x) \tag{121}$$

where $\mathbf{h}_k$ is also a possibly nonlinear function of the state and the measurement noise $\mathbf{v}_k$. We assume the process noise and the measurement noise are white and statistically independent. The second criterion is very difficult to prove, in which case, for practical purposes we accept that they are at least uncorrelated. Succinctly,

$$E\{\mathbf{w}_i\mathbf{w}_j^T\} = E\{\mathbf{v}_i\mathbf{v}_j^T\} = \begin{cases} \delta(0), \ \forall i = j \\ 0, \quad \forall i \neq j \end{cases} \tag{122}$$

$$E\{\mathbf{w}_i\mathbf{v}_j^T\} = 0, \forall i, j \tag{123}$$

where $E\{\cdot\}$ is the expectation operator and $\delta(\cdot)$ is the Dirac delta function. At any point, $\mathbf{x}$ will be a random sample associated with a particular probability density function (pdf). More specifically, given all the measurements of the system up to time $k-1$, we will have the conditional pdf $p(\mathbf{x}_k|\mathbf{Z}_{k-1})$ where $\mathbf{Z}_{k-1} = \{\mathbf{z}_1, \mathbf{z}_2, \ldots, \mathbf{z}_{k-1}\}$. The goal is to find $p(\mathbf{x}_k|\mathbf{Z}_k)$ once new measurements are available. Because the current state is dependent only on the state immediately preceding it, it is first-order Markovian and we apply the Chapman-Kolmogorov equation (Duda et al., 2001; Ristic et al., 2004):

$$p(\mathbf{x}_k|\mathbf{Z}_{k-1}) = \int p(\mathbf{x}_k|\mathbf{x}_{k-1})p(\mathbf{x}_{k-1}|\mathbf{Z}_{k-1})d\mathbf{x}_{k-1} \tag{124}$$

where $p(\mathbf{x}_k|\mathbf{x}_{k-1})$ is the transition density, which allows us to calculate the probability that a state will evolve in a particular way from one instant to the next. The result of (124) is essentially a *prediction* of the state vector given all previous information (the Bayesian prior pdf) . Once new measurements become available, we seek to *update* the estimate of $\mathbf{x}_k$ using $\mathbf{z}_k$ (or obtain the Bayesian posterior pdf). Using Bayes' formula

$$\begin{aligned} p(\mathbf{x}_k|\mathbf{Z}_k) &= \frac{p(\mathbf{z}_k|\mathbf{x}_k)p(\mathbf{x}_k|\mathbf{Z}_{k-1})}{p(\mathbf{z}_k|\mathbf{Z}_{k-1})} \\ &= \frac{p(\mathbf{z}_k|\mathbf{x}_k)p(\mathbf{x}_k|\mathbf{Z}_{k-1})}{\int p(\mathbf{z}_k|\mathbf{x}_k)p(\mathbf{x}_k|\mathbf{Z}_{k-1})d\mathbf{x}_k} \end{aligned} \tag{125}$$

We can recursively employ (124) and (125) to estimate the state pdf at any time, from which estimates of the state vector itself can be obtained using any optimality criterion we chose.

Initially, there will, of course, be no prior states or measurements, so we define the initial prior as $p(\mathbf{x}_0)$, which is simply our best estimate given what we know generally about such systems.

5.2 Linear filters

Linear filtering attempts to find the optimal linear combination of the predicted state and the the state implied by the measurements. In recursive form, a linear filter can be written as

$$\hat{\mathbf{x}}_{k|k} = \mathbf{K}'_k \hat{\mathbf{x}}_{k|k-1} + \mathbf{K}_k \mathbf{z}_k \tag{126}$$

where $\mathbf{K}'_k$ and $\mathbf{K}_k$ are weight or gain matrices to be computed at each instant k. Under the assumption of Gaussian noise, the resulting pdf of the estimate will also be Gaussian. The Kalman filter, which we shall describe next, is the optimal estimator under these circumstances (Kalman, 1960).

5.3 The Kalman filter

The problem as described in its general form in section 5.1 is rarely analytically tractable in practice. For one thing, state vectors may be very large, having high- (or infinite-) dimensional pdfs which cannot be integrated in the denominator of (125); the well-known "curse of dimensionality"(Duda et al., 2001). Secondly, the nonlinear models themselves may be unavailable or too complex to deal with analytically. For this reason, the error dynamics in 3.6 have been approximated to a linear system. We have also made many linearizing assumptions in the state equations such as the assumption of fast sampling rates relative to vehicle dynamics. We have also made the assumption of Gaussian-distributed noise in the sensors and shall further assume that disturbances to the error dynamics model take the same (though uncorrelated) form.

Under these conditions, that is linearity and Gaussianity, a realizable solution to the Bayesian estimation problem is the Kalman filter, first described by Rudolph Kalman in 1960. Let the discrete-time system and its measurement equation be defined as

$$\mathbf{x}_k = \mathbf{\Phi}_{k,k-1}\mathbf{x}_{k-1} + \mathbf{G}_{k-1}\mathbf{w}_{k-1} \tag{127}$$

$$\mathbf{z}_k = \mathbf{H}_k\mathbf{x}_k + \mathbf{v}_k \tag{128}$$

where $\mathbf{\Phi}_{k,k-1}$ is the state transition matrix, $\mathbf{w}_{k-1}$ is the zero-mean process noise with shaping matrix $\mathbf{G}_{k-1}$, $\mathbf{H}_k$ is the design matrix mapping state parameters to measurements in $\mathbf{z}_k$ and $\mathbf{v}_k$ is zero-mean measurement noise (uncorrelated with process noise). $\mathbf{\Phi}_{k,k-1}$ and $\mathbf{H}$ are linear functions, so $\mathbf{x}_k$ and $\mathbf{z}_k$ are Gaussian random vectors whose pdfs are completely described by their means and covariances. In addition to the requirements of (122) and (123), we have for the initial state $\mathbf{x}_0$ the following constraints:

$$E\{\mathbf{x}_0\mathbf{w}_k^T\} = 0 \tag{129}$$

$$E\{\mathbf{x}_0\mathbf{v}_k^T\} = 0 \tag{130}$$

We seek an efficient estimator (if it exists), which is to say it is unbiased and its covariance attains the Cramér-Rao Lower Bound (CRLB). Such an estimator, if it exists, is both the minimum variance unbiased estimator (MVUE) and the maximum-likelihood estimator

(MLE), thought the converse is generally not true. An efficient estimator therefore has the properties that its estimates $\hat{\mathbf{x}}_k$ and those of any other estimator $\tilde{\mathbf{x}}_k$ satisfy

$$E\{(\hat{\mathbf{x}}_k - \mathbf{x}_k)^T(\hat{\mathbf{x}}_k - \mathbf{x}_k)\} \leq E\{(\tilde{\mathbf{x}}_k - \mathbf{x}_k)^T(\tilde{\mathbf{x}}_k - \mathbf{x}_k)\} \tag{131}$$

where (in the case that both are unbiased),

$$E\{\hat{\mathbf{x}}_k\} = E\{\tilde{\mathbf{x}}_k\} = E\{\mathbf{x}_k\} \tag{132}$$

The Gaussian likelihood function has the form

$$\ell(\boldsymbol{\mu}, \boldsymbol{\Sigma}; \mathbf{x}) = \exp(-\frac{1}{2}(\mathbf{x} - \boldsymbol{\mu})^T \boldsymbol{\Sigma}^{-1}(\mathbf{x}-\boldsymbol{\mu})) \tag{133}$$

where the free parameters are the mean $\boldsymbol{\mu}$ and the covariance $\boldsymbol{\Sigma}$ and the data $\mathbf{x}$ is given. A maximum likelihood estimator is one that provides $\boldsymbol{\mu}$ and $\boldsymbol{\Sigma}$ that maximize ℓ given a particular sample $\mathbf{x}$. In other words, it provides the parameters of the Gaussian pdf that would most likely lead to having observed $\mathbf{x}$ (Papoulis, 1977). For two independent events, the joint probability is $P(a, b) = P(a)P(b)$. Likelihood functions obey a similar rule

$$\ell_{A,B}(\boldsymbol{\mu}_{A,B}, \boldsymbol{\Sigma}_{A,B}; \mathbf{x}) = \ell_A(\boldsymbol{\mu}_A, \boldsymbol{\Sigma}_A; \mathbf{x})\ell_B(\boldsymbol{\mu}_B, \boldsymbol{\Sigma}_B; \mathbf{x}) \tag{134}$$

which, for the log-likelihood reduces to the equality

$$\log(\ell_{A,B}(\boldsymbol{\mu}_{A,B}, \boldsymbol{\Sigma}_{A,B}|\mathbf{x}, \mathbf{y})) = \log(\ell_A(\boldsymbol{\mu}_A, \boldsymbol{\Sigma}_A|\mathbf{x})) + \log(\ell_B(\boldsymbol{\mu}_B, \boldsymbol{\Sigma}_B|\mathbf{x}))$$

or

$$\begin{aligned} &-\frac{1}{2}(\mathbf{x} - \boldsymbol{\mu}_{A,B})^T \boldsymbol{\Sigma}_{A,B}^{-1}(\mathbf{x}-\boldsymbol{\mu}_{A,B}) = \\ &\log(c) - \frac{1}{2}(\mathbf{x} - \boldsymbol{\mu}_A)^T \boldsymbol{\Sigma}_A^{-1}(\mathbf{x}-\boldsymbol{\mu}_A) - \frac{1}{2}(\mathbf{x} - \boldsymbol{\mu}_B)^T \boldsymbol{\Sigma}_B^{-1}(\mathbf{x}-\boldsymbol{\mu}_B) \end{aligned} \tag{135}$$

where $c > 0$ is an arbitrary constant. As shown in (Grewal et al., 2001), after taking the first and second derivatives with respect to $\mathbf{x}$, this further reduces to

$$\boldsymbol{\Sigma}_{A,B}^{-1} = \boldsymbol{\Sigma}_A^{-1} + \boldsymbol{\Sigma}_B^{-1} \tag{136}$$

Furthermore, the joint MLE is given by

$$\boldsymbol{\Sigma}_{A,B}^{-1}\boldsymbol{\mu}_{A,B} = \boldsymbol{\Sigma}_A^{-1}\boldsymbol{\mu}_A + \boldsymbol{\Sigma}_B^{-1}\boldsymbol{\mu}_B \tag{137}$$

$$\Rightarrow \boldsymbol{\mu}_{A,B} = (\boldsymbol{\Sigma}_A^{-1} + \boldsymbol{\Sigma}_B^{-1})^{\dagger}(\boldsymbol{\Sigma}_A^{-1}\boldsymbol{\mu}_A + \boldsymbol{\Sigma}_B^{-1}\boldsymbol{\mu}_B) \tag{138}$$

where $\dagger$ is the generalized inverse of the matrix.

For any measurement vector $\mathbf{z}$, its ML estimate $\hat{\boldsymbol{\mu}}_{\mathbf{z}} = \mathbf{z}$ and its covariance $\boldsymbol{\Sigma}_{\mathbf{z}} = E\{\mathbf{v}\mathbf{v}^T\}$. We now transform the likelihood based on the measurements into likelihood of the state vector by

$$\boldsymbol{\mu}_{\mathbf{x}} = \mathbf{H}^{\dagger}\mathbf{z} \tag{139}$$

$$\boldsymbol{\Sigma}_{\mathbf{x}}^{-1} = \mathbf{H}^T\boldsymbol{\Sigma}_{\mathbf{z}}^{-1}\mathbf{H} \tag{140}$$

Let $\boldsymbol{\mu}_A = \hat{\mathbf{x}}_{k|k-1}$, that is the MLE prior to update, and $\boldsymbol{\Sigma}_A^{-1} = \mathbf{P}_{k|k-1}^{-1}$ or the inverse of covariance matrix of the MLE prior to update, whose propagation in time is given by

$$\hat{\mathbf{x}}_{k|k-1} = \boldsymbol{\Phi}_{k,k-1}\hat{\mathbf{x}}_{k-1|k-1} \tag{141}$$

and

$$\mathbf{P}_{k|k-1} = \boldsymbol{\Phi}_{k,k-1}\mathbf{P}_{k-1|k-1}\boldsymbol{\Phi}_{k,k-1}^T + \mathbf{Q}_{k-1} \tag{142}$$

where $\mathbf{Q}_{k-1}$ is the covariance of the process noise $\mathbf{w}_{k-1}$. Next, let $\boldsymbol{\mu}_B = \boldsymbol{\mu}_{\mathbf{x}}$ as defined in (139) and $\boldsymbol{\Sigma}_B^{-1} = \boldsymbol{\Sigma}_{\mathbf{x}}^{-1}$ as defined in (140). Now, the covariance after the update $\mathbf{P}_{k|k}$ is

$$\begin{aligned}
\mathbf{P}_{k|k} &= \boldsymbol{\Sigma}_{A,B} \\
&= (\boldsymbol{\Sigma}_A^{-1} + \mathbf{H}^T\boldsymbol{\Sigma}_{\mathbf{z}}^{-1}\mathbf{H})^{-1} \\
&= \boldsymbol{\Sigma}_A - \boldsymbol{\Sigma}_A\mathbf{H}^T(\mathbf{H}\boldsymbol{\Sigma}_A\mathbf{H}^T + \boldsymbol{\Sigma}_{\mathbf{z}})^{-1}\mathbf{H}\boldsymbol{\Sigma}_A \\
&= \mathbf{P}_{k|k-1} - \mathbf{P}_{k|k-1}\mathbf{H}^T(\mathbf{H}\mathbf{P}_{k|k-1}\mathbf{H}^T + \boldsymbol{\Sigma}_{\mathbf{z}})^{-1}\mathbf{H}\mathbf{P}_{k|k-1}
\end{aligned} \tag{143}$$

The estimate after the update is $\hat{\mathbf{x}}_{k|k} = \boldsymbol{\mu}_{A,B}$, which by (138), is

$$\begin{aligned}
\hat{\mathbf{x}}_{k|k} &= (\boldsymbol{\Sigma}_A^{-1} + \boldsymbol{\Sigma}_B^{-1})^{\dagger}(\boldsymbol{\Sigma}_A^{-1}\boldsymbol{\mu}_A + \boldsymbol{\Sigma}_B^{-1}\boldsymbol{\mu}_B) \\
&= \mathbf{P}_{k|k}(\mathbf{P}_{k|k-1}\hat{\mathbf{x}}_{k|k-1} + \mathbf{H}^T\boldsymbol{\Sigma}_{\mathbf{z}}^{-1}\mathbf{H}\mathbf{H}^{\dagger}\mathbf{z})
\end{aligned}$$

which, after some simplification, becomes

$$\begin{aligned}
\hat{\mathbf{x}}_{k|k} &= \hat{\mathbf{x}}_{k|k-1} + \mathbf{P}_{k|k-1}\mathbf{H}^T(\mathbf{H}\mathbf{P}_{k|k-1}\mathbf{H}^T + \boldsymbol{\Sigma}_{\mathbf{z}})^{-1}(\mathbf{z} - \mathbf{H}\hat{\mathbf{x}}_{k|k-1}) \\
&= \hat{\mathbf{x}}_{k|k-1} + \mathbf{K}(\mathbf{z} - \mathbf{H}\hat{\mathbf{x}}_{k|k-1})
\end{aligned} \tag{144}$$

Where $\mathbf{K}$ is the Kalman gain matrix, which provides the optimal weights for combining the predicted estimate of the state with the new measurements. After the update, the current estimate and its covariance is propagated in time using (141) and (142)

It can be shown that the updated state covariance matrix achieves the Cramér-Rao Lower Bound, and that $\hat{\mathbf{x}}_{k|k}$ is unbiased, meaning that the estimator is efficient and automatically both the MLE and the MVUE. Finally, in terms of pdfs,

$$p(\mathbf{x}_{k-1}|\mathbf{Z}_{k-1}) = \mathcal{N}(\mathbf{x}_{k-1}; \hat{\mathbf{x}}_{k-1|k-1}, \mathbf{P}_{k-1|k-1}) \tag{145}$$

$$p(\mathbf{x}_k|\mathbf{Z}_{k-1}) = \mathcal{N}(\mathbf{x}_k; \hat{\mathbf{x}}_{k|k-1}, \mathbf{P}_{k|k-1}) \tag{146}$$

$$p(\mathbf{x}_k|\mathbf{Z}_k) = \mathcal{N}(\mathbf{x}_k; \hat{\mathbf{x}}_{k|k}, \mathbf{P}_{k|k}) \tag{147}$$

which provides a Bayesian interpretation of the Kalman filter.

5.4 GNSS aiding

Because of unbounded position errors associated with misalignment and gyro drift, along with the undesirability of having even bounded oscillations in the position due to accelerometer and velocity errors, it is necessary for most applications using medium-grade and commodity-grade IMUs to employ an aiding method. That is, using an external (and independent) estimate of navigation states to limit the accumulation of errors in the INS. For the last two decades, the preferred method has been to use measurements obtained from global navigation satellite systems such as GPS to update the INS error estimates and improve the navigation solution. The simplest way to achieve this, of course, is to simply use the calculated positions and velocities from the GNSS directly in place of the results from the mechanization as they become available. This is the so-called reset "filter", although from the standpoint of optimal filtering, it has many undesirable effects such as introducing sudden jumps in the navigation states. Moreover, the complementary filter places all the weight on the GNSS-derived values, which themselves are subject to error.

Alternatively, Kalman filtering is used to optimally estimate the error states of the INS, with updates coming from GNSS in one of several architectures:

- Loosely-coupled integration. Here, the GNSS system acts to provide a full position and velocity estimate independently of the INS mechanization. The measurements of the error states arises from subtracting the GNSS states from the position and velocity arising from the mechanization. These are then transferred through the design matrix **H** of the measurement equations and used to update the error estimates, which in turn are subtracted from the navigation states.
- Tightly-coupled integration. In this scheme, the GNSS measurements and error states are directly incorporated into the Kalman filter, the primary benefit being that the navigation state can be improved over the mechanization alone with fewer than four GNSS satellites being tracked at any given time. A detailed treatment can be found in (Grewal et al., 2001).
- Deeply-coupled integration. This is a hardware-level implementation which further incorporates the states associated with the GNSS receiver signal tracking loop. This allows for better tracking stability under high dynamics and rapid reacquisition of GNSS signals under intermittent visibility (Kim et al., 2006; Kreye et al., 2000).

In the loosely-coupled scheme, at each update we let

$$\mathbf{z} = \begin{pmatrix} \phi_{INS} - \phi_{GNSS} \\ \lambda_{INS} - \lambda_{GNSS} \\ h_{INS} - h_{GNSS} \\ v_{e_{INS}} - v_{e_{GNSS}} \\ v_{n_{INS}} - v_{n_{GNSS}} \\ v_{u_{INS}} - v_{u_{GNSS}} \end{pmatrix} = \begin{pmatrix} \delta\phi \\ \delta\lambda \\ \delta h \\ \delta v_e \\ \delta v_n \\ \delta v_u \end{pmatrix} \tag{148}$$

The measurement equation is then

$$\mathbf{H}\hat{\mathbf{x}}_{k|k-1} = \begin{pmatrix} 1\,0\,0\,0\,0\,0\,0\,0\,0\,0\,0\,0\,0\,0\,0 \\ 0\,1\,0\,0\,0\,0\,0\,0\,0\,0\,0\,0\,0\,0\,0 \\ 0\,0\,1\,0\,0\,0\,0\,0\,0\,0\,0\,0\,0\,0\,0 \\ 0\,0\,0\,1\,0\,0\,0\,0\,0\,0\,0\,0\,0\,0\,0 \\ 0\,0\,0\,0\,1\,0\,0\,0\,0\,0\,0\,0\,0\,0\,0 \\ 0\,0\,0\,0\,0\,1\,0\,0\,0\,0\,0\,0\,0\,0\,0 \end{pmatrix} \hat{\mathbf{x}}_{k|k-1} \tag{149}$$

Under the assumption of a fast sampling rate, the state transition matrix $\mathbf{\Phi}_{k,k-1}$ is given by

$$\mathbf{\Phi}_{k,k-1} = e^{\delta t \mathbf{F}} \tag{150}$$

Where **F** is the error dynamics matrix defined in 3.6. Using (141),(142), (143) and (144), it becomes evident that although alignment errors and sensor biases are not directly observable, the Kalman gain matrix **K** contains information about their contribution to the navigation states. Since the error dynamics are linearized about the current state estimates, the filter presented here is an example of extended Kalman filtering.

We finish with the remark that, in general, the assumptions of linearity, Gaussianity and uncorrelated noise sources are not strictly justified in INS applications. Beyond employing an extended Kalman filter (EKF) as described above, the further development and application of nonlinear filters such as the unscented Kalman filter (UKF) and the particle filter have been undertaken in attempts at solving the more general Bayesian formulation of the problem.

6. Conclusion

Aided inertial navigation remains an active area of research, especially with the introduction of smaller and cheaper (but noisier) inertial sensors. Among the challenges presented by these devices is heading initialization (Titterton & Weston, 2004), which necessitates the use of other aiding systems, and proper stochastic modeling of their error charactertics. In addition, the nonlinearity of the state equations has prompted much research in applied optimal estimation. Despite this, the underlying concepts remain the same and the development presented here should give the reader enough background to understand the issues involved, enabling him or her to pursue more detailed aspects of INS and aided INS design as necessary.

7. References

Angeles, J.(2003). *Fundamentals of robotic mechanical systems: theory, methods, and algorithms*, Springer-Verlag, New York.

Duda, R.; Hart, P. & Stork, D.(2001). *Pattern Classification*, John Wiley & Sons, New York.

Featherstone, W.E. & Dentith, M.C. (1997). A Geodetic Approach to Gravity Data Reduction for Geophysics. *Computers and Geosciences*, 23, 10, 1063-1070

Grewal, M.S.; Weill, L.R. & Andrews, A.P.(2001). *Globabl Positioning Systems, Inertial Navigation, and Integration*, John Wiley & Sons, New York.

Hofmann-Wellenhof; B. Lichtenegger, H. & Collins, J. (2001). *Global Positioning System Theory and Practice*, Springer-Verlag, Wien and New York.

Hofmann-Wellenhof, B. & Moritz, H. (2005).*Physical Geodesy, Second Edition*, Springer-Verlag, Wien and New York.

Kalman, R.E. (1960). A New Approach to Linear Filtering and Prediction Problems. *Transactions of the ASME Journal of Basic Engineering*, 82, 34-45, 1960.

Kim, J.W.; Hwang, D. & Lee, S.J. (2006). A Deeply Coupled GPS/INS Integrated Kalman Filter Design Using a Linearized Correlator Output. *Position, Location, And Navigation Symposium, 2006 IEEE/ION* , pp. 300-305, April 2006, IEEE, San Diego.

Kohler, W.E. & Johnson, L. (2006). *Elementary Differential Equations with Boundary Value Problems*, Pearson/Addison Wesley, Boston.

Kreye, C.; Eissfeller, B. & Winkler, J.O. (2000). Improvement of GNSS Receiver Performance Using Deeply Coupled INS Measurement. *Proceedings of ION GPS*, pp. 844-854, The Institute of Navigation, Salt Lake City.

Kuipers, J. (1999). *Quaternions and Rotation Sequences*, Princeton University Press, Princeton.
NIMA (2000). *Department of Defense World Geodetic System 1984*, National Imagery and Mapping Agency, NIMA Stock Number DMATR83502WGS84.
Pio, R.L. (1966). Euler Angle Transformations. *IEEE Transactions on Automatic Control*, AC-11, 4, 1966.
Papoulis, A. *Signal Analysis*, McGraw-Hill, New York.
Ristic, B.; Arulampalam, S. & Gordon, N. (2004). *Beyond the Kalman Filter: Particle Filters for Tracking Applications*, Artech House, Boston.
Rogers, R. (2003). *Applied Mathematics in Integrated Navigation Systems, Second Edition*, American Institute of Aeronautics and Astronautics, Reston.
Schwarz, K.P. & Wei, M. (1990). Efficient Numerical Formulas for the Computation of Normal Gravity in a Cartesian Frame. *Manuscripta Geodetica*, 4, 15, 228-234.
Titterton, D. & Weston, J. (2004). *Strapdown Inertial Navigation Technology, 2nd Edition*, The Institution of Electrical Engineers, Herts.
Torge, W. (2001). *Geodesy, 3rd Edition*, Walter de Gruyter, Berlin and New York.

2

Gain Tuning of Flight Control Laws for Satisfying Trajectory Tracking Requirements

Urbano Tancredi[1] and Federico Corraro[2]
[1]*University of Naples Parthenope*
[2]*Italian Aerospace Research Centre*
Italy

1. Introduction

The present chapter is concerned with presenting an approach for the synthesis of a gain-scheduled flight control law that assures compliance to trajectory tracking requirements. More precisely, a strategy is proposed for improving the tracking performances of a baseline controller, obtained by conventional synthesis techniques, by tuning its gains. The approach is specifically designed for atmospheric re-entry applications, in which gain scheduled flight control laws are typically used.

Gain-scheduling design approaches conventionally construct a nonlinear controller by combining the members of an appropriate family of linear time-invariant (LTI) controllers (Leith & Leithead, 2000). The time-invariant feedback laws usually share the same structure, and differ only for the values of some tunable parameters, most notably the controller's gains. These gains are generally determined taking advantage of well-assessed LTI-based design techniques, such as pole placement and gain/phase margin methods. However, once a set of LTI feedback laws is specified, the nonlinear controller must be synthesized, which requires an additional design step. This step is of considerable importance since the choice of nonlinear controller realization can greatly influence the closed loop performance (Leith & Leithead, 2000). Furthermore, actual mission requirements constraint quantitatively the time response of the augmented system (Crespo et al., 2010), e.g. by imposing tracking requirements of a reference trajectory or requiring relevant output variables to be enclosed within a limited flight envelope. As such, the final gain-scheduled controller's performances are ascertained by means of numerical simulation based methods, most notably Monte Carlo, which can highlight limitations that were not apparent in the LTI design phase. As a result, in these cases one is forced to iterate the LTI design, but using analysis results that refer to the nonlinear controller rather than to the LTI ones, further complicating the design improvement task.

Several methods have been proposed in the open literature both for taking into account explicitly the complex dependency of the final controller response from its gains and for dealing with quantitative performance requirements, such as tracking errors. Most, if not all, proposed approaches formulate the design task as an optimization problem, in which the merit function evaluation requires numerical simulation of the augmented system's time-response. For instance, (Crespo et al., 2008) develops optimization-based strategies for control analysis and tuning at the control verification stage, which build upon numerical evaluation of controller's performance metrics that require simulation of the augmented model. Other authors (dos Santos Coelho, 2009) suggest using chaotic optimization algorithms for enhancing the computational efficiency of the numerical optimization problem. In (Wang & Stengel, 2002), a robust control law is synthesized using probabilistic robustness techniques, by minimizing a cost that is a function of the probabilities that design criteria will not be satisfied. Monte Carlo simulation is used to estimate the likelihood of system instability and violation of performance requirements subject to variations of the probabilistic system parameters. Stochastic parameter tuning is also proposed in (Miyazawa & Motoda, 2001), which is a form of optimization by which the probability of the total mission achievement is maximized w.r.t the flight control system's tunable parameters. Mission achievement probability is estimated by applying the Monte Carlo method also in this case.

In this chapter, we propose a methodology for determining all combinations, within a given domain, of the flight control law tunable gains that comply with quantitative requirements expressed in the time domain, and is applicable to nonlinear control laws such as gain-scheduled flight control ones. This approach aims at providing quantitative indications on the Flight Control Law (FCL) time-domain performance, taking explicitly into account the complex dependency given by the scheduling of the LTI control laws. As such, it is intended to complement the conventional LTI-based controller synthesis approaches, such as pole/placement and frequency domain methods, which are thus still in charge of addressing the system's stability and robustness.

The approach is based on a technique developed by the authors for tackling a different problem, namely the robustness analysis of a given flight control law (Tancredi et al., 2009). It builds upon a Practical Stability criterion, in which the allowable trajectories dispersion can be specified in the time-domain, in an extremely appealing manner to enforce practical engineering requirements. Under the assumption that the gains domain is a convex polytope, the method results allow distinguishing in the whole domain the gain combinations matching the criterion from those yielding unsatisfactory performance. This is done inferring the nonlinear augmented system behaviour for all gains ranging in a convex polytope from numerical simulations of the augmented dynamics at a limited number of specific points of the gains domain. A set inversion algorithm selects these points using an adaptive gridding strategy. The proposed technique is applied to a gain-scheduled flight control law of the Unmanned Space Vehicle, a re-entry technology demonstrator pursued by the Italian Aerospace Research Centre. Results demonstrate the method's effectiveness in determining the gains combinations allowing to satisfy pre-specified trajectory tracking requirements. Results also show that it is computationally viable and that it allows gaining insight into the factors that limit the controller's performance, thus aiding eventual additional LTI-based design iterations.

2. Problem setting

We refer in this work to atmospheric re-entry applications, and to a FCL whose gains are scheduled depending on the values of some specifically selected independent variables, either being a univocal function of the system dynamical state vector, such as air-relative velocity, altitude, Mach number and so on, or explicitly dependent on time. Selection of the scheduling law and of the independent variables is out of the scope of the present chapter, because it typically involves exploiting the peculiar flight mechanics features of the application at hand. We assume henceforth that the FCL structure is known, and that the FCL is completely specified once a limited number of parameters, i.e. the FCL gains, are set to a constant value. As introduced in the previous section, the problem dealt with in this chapter is to determine the values of these gains that allow complying to trajectory tracking requirements. Let us assume to have a starting design point that specifies a set of gain values, which typically does not allow satisfying the tracking requirements. We denote this initial guess as the nominal gain value, which is taken equal to zero to simplify notation. Let us also assume to have a finite number p of constant gains and that the gains are enclosed in a bounded set $\Pi \subseteq \Re^p$, which represents the region in the gains space one wishes to analyze.

The dynamical system we refer to shall be suitable to represent the closed-loop augmented dynamics of an atmospheric re-entry vehicle. The typical FCL for this application foresee a gain-scheduled inner-loop PID control scheme coupled with a time-varying guidance law, possibly dependent on the system state as well. Gain scheduling is taken into account by dependency on the state variables (and time if needed), and the PID action by dependencies on the state, on its time integral (which adds up to the open-loop system's state) and derivative, respectively. Thus, let us consider the following dynamical system, in which $x \in \Re^n$, $y \in \Re^w$, and the feedback action is included in the $f(\cdot)$ and $g(\cdot)$ functions.

$$\dot{x} = f(t, x, \pi) \quad y = g(t, x, \pi) \tag{1}$$

In the case of an un-powered re-entry vehicle in steep gliding flight, due to the lack of stationary equilibrium solutions for Eq. (1), we refer to time-varying nominal trajectories rather than stationary operating conditions. In addition, these trajectories are usually defined on a finite-time domain, i.e. $t \in [0,T]$, where the initial epoch is taken equal to zero for simplicity and T is a finite positive real number. The nominal trajectory is thus time varying on a compact time domain, and satisfies the following equations.

$$\tilde{x}(t) := \tilde{x}_0 + \int_0^t f(\tau, \tilde{x}, 0) \times d\tau \quad \forall t \in [0,T] \tag{2a}$$

$$\tilde{y}(t) := g(t, \tilde{x}, 0) \quad \forall t \in [0,T] \tag{2b}$$

For gain-scheduled FCL, the current design practice relies on the well-known theory of Linear Time Invariant (LTI) systems. In this approach, the original nonlinear system representing the augmented vehicle dynamics is linearized around a limited number of representative time-varying trajectories, including the nominal one. Then, the well-known frozen-time approach is applied (Lee & Choi, 2004), yielding multiple LTI models. In this way classical design techniques, such as pole placement and gain/phase margin methods, can be exploited. Even if the flight experience has demonstrated that this approach is indeed operative, it is also widely recognized as inefficient (Leith & Leithead, 2000). In fact, LTI-based analysis may call for gain design values for counteracting poor closed-loop dynamic performances (for instance, low damping or even instabilities) in some of the chosen points on the trajectories. However, the effect of undesirable frozen-time performances on the overall mission objectives can be of scarce importance since the vehicle remains in a particular frozen time condition only for a limited amount of time. Thus, modification of the

FCL for improving the LTI-based dynamic performances could be un-necessary, since these missions typically specify time-domain criteria, such as nominal trajectory tracking performances, which can be satisfied also in presence of poor frozen-time dynamic performances. LTI-based analysis results are thus usually complemented by dedicated numerical-simulation based analyses, such as Monte Carlo techniques, through which the quantitative dispersion about the reference trajectory can be estimated. Finally, in the LTI-based approach, the gain tuning problem shall be solved in each frozen operating condition, thus considerably limiting the dimension of manageable problems.

The criterion proposed in the present work is instead based on the Practical Stability and/or Finite-Time Stability concepts, whose detailed description can be found in (Gruyitch et al., 2000; Dorato, 2006). This type of stability requires only the inclusion of the system trajectories in a pre-specified subset of the state space, possibly time-varying, in face of bounded initial state displacements and disturbances. As opposed to the classical Lyapunov stability concept it does not require the existence of any equilibrium point, and is independent from Lyapunov stability, in the sense that one neither implies nor excludes the other. The practical stability criterion is inherently well suited to the applications of interest: it allows to take explicitly into account system (1) time domain finiteness, and to use criteria directly linked to the original mission or system requirements, which are typically expressed in terms of trajectory tracking performances. Indeed, the latter can be easily enforced by requiring the inclusion of the system trajectories in a pre-specified time-varying subset of the state space determined by the tracking requirements, to which we refer as the admissible solutions tube, $S_A(t)$.

Let us assume the initial state to be perfectly known and equal to the nominal one. In other words, the perturbed output trajectory $y(t;\pi)$ is defined as a trajectory of system (1) that starts at $t = 0$ in $y(0) = y_0$, under the constant input π. This assumption does not limit the scope of the problem, since initial state dispersions can be included, if necessary, as additional elements of the π vector with no conceptual modifications. The tracking requirements are used to define a Boolean property P depending on the gains, so that the system complies with the practical stability criterion if and only if the property is true. In order to gain generality in the capability to enforce admissible dispersion requirements, P is defined in terms of the output trajectories of system (1) (that cover the case in which the system state is analyzed by letting $y = x$).

$$P(\pi) = \begin{cases} true & y(t;\pi) \in S_A(t) \ \ \forall t \in [0,T] \\ false & \exists t \in [0,T] : y(t;\pi) \notin S_A(t) \end{cases} \tag{3}$$

The capability of identifying all the combinations of the gain values in Π for which the property is true can greatly aid the refinement of the candidate FCL design. Indeed, it allows analyzing the FCL performance over the whole gain domain, as opposed to classical analysis that identifies only a limited number of points in Π. This aids the FCL upgrade by simplifying the physical understanding of the causes for poor performance. This feature is highly desirable in a design context, in which rather than the knowledge of a certain requirement violation is the determination of the causes that mainly contributes to identify possible design refinement strategies. With this in mind, the gain tuning task is stated as determining the set Π_A, subset of Π, which is made of all the *admissible* gains, that is, all gains satisfying the tracking requirements.

$$\Pi_A := \{\pi \in \Pi \mid P(\pi) = true\} \tag{4}$$

In this setting, the gain tuning task can be re-formulated as a practical stability analysis problem, as follows.

Problem 1. Given system (1), a bounded set $\Pi \subseteq \Re^p$ such that $\pi \in \Pi$, a time-varying compact set $S_A(t)$ (admissible solutions tube), and the property P, determine the set Π_A.

3. Solution approach

In order to simplify the solution to problem 1, we introduce the following restricting assumptions.

Assumption 1. The functions $f(\cdot)$ and $g(\cdot)$ are differentiable in t, x and π over relevant domains.

Assumption 2. The gains range in a p–dimensional hyper-rectangle Π.

$$\Pi := [\underline{\pi}_1, \overline{\pi}_1] \times .. \times [\underline{\pi}_p, \overline{\pi}_p] \tag{5}$$

Assumption 3. The required solutions tube is a w–dimensional hyper-rectangle for all $t \in [0,T]$

$$S_A(t) := [\underline{S}_{1A}(t), \overline{S}_{1A}(t)] \times .. \times [\underline{S}_{wA}(t), \overline{S}_{wA}(t)] \tag{6}$$

Various techniques exist being able to deal with the practical stability analysis of a nonlinear dynamical system (see Dorato, 2006, for a survey). The prominent approaches are based on a Lyapunov-type analysis involving an auxiliary function referred to as a Lyapunov-like function in (Gruyitch et al., 2000; Dorato, 2006). However, to the authors' knowledge, there are no systematic and operative means to find a suitable Lyapunov function when nonlinear time-varying systems are considered; Lyapunov-based methods are also inherently conservative in estimating the trajectories dispersion, depending on the selected Lyapunov-like function. A different approach is presented in (Ryali & Moudgalya, 2005), which stems from the notion of positively invariant tubes. However, it does not bound nor estimates the results conservativeness, with a resulting limited applicability to problems of practical interest. Finally, for Linear Time-Varying (LTV) systems, practical stability analysis approaches have been developed based on operator theory (Amato et al., 2003), which yield only sufficient conditions in the form of a nonlinear, time-varying, differential matrix inequality. Generally speaking, in spite of a wide literature on practical stability theoretical results, all the reported approaches suffer of significant drawbacks when considered from an applicability perspective, including cases where the system dynamics are linear. Indeed, the abundance of theoretical results on practical stability analysis methods it is not balanced by examples of their application to cases of practical engineering interest within the robustness analysis context.

The approach followed in this chapter extends the one proposed in (Tancredi et al., 2009). for analyzing the robustness of a given flight control law. By setting up the gain tuning task as in Problem 1, this approach can be adapted for being used with the problem at hand with only minor modifications. An overview of the method is repeated in this chapter closely following the one in (Tancredi et al., 2009), but providing additional details and adapting it for dealing with a gain tuning problem. The technique approximates the solution of the practical stability analysis problem for a complex system with the solutions obtained for simpler systems, for which an efficient solution approach can be found. Specifically, the proposed solution approach foresees two successive phases. First, the nonlinear vehicle dynamics are approximated within a pre-specified error tolerance by their time-varying linearizations under several off-nominal gains (approximation phase). Then, problem 1 is solved on the LTV systems obtained in the previous phase taking explicitly into account the approximation error. This is done performing numerical simulations only at suitably selected gains combinations and exploiting the convexity preservation property of the LTV dynamics (property clearance phase). For the sake of clarity, we will describe separately these two phases.

3.1 Approximation

Let us consider a partition $\{\Pi_k\}$ of the gain domain, made of hyper-rectangular blocks Π_k, that is, a collection of subsets (blocks) that are both collectively exhaustive and mutually exclusive with respect to the set being partitioned. We then define a collection of LTV systems, each one approximating the nonlinear system in a single block. In particular, each LTV system is obtained linearizing the system around its trajectory obtained by setting the uncertainties to π_k^0, the geometrical centre of Π_k. The dynamic equations for each one of such LTV systems as π ranges in the relevant Π_k can be written as:

$$\dot{x}_{Lk} = \dot{x}_k^0 + A_k \times \left(x_{Lk} - x_k^0\right) + G_k \times \left(\pi - \pi_k^0\right) \tag{7a}$$

$$y_{Lk} = y_k^0 + C_k \times \left(x_{Lk} - x_k^0\right) + D_k \times \left(\pi - \pi_k^0\right) \tag{7b}$$

where the A_k, G_k, C_k, and D_k matrixes are obtained applying first order expansion of the nonlinear functions in Eq. (1) around x_k^0, π_k^0. Note that, being the nonlinear system's trajectories time-varying, the centre trajectory and the matrices in Eq. (7) are in general time-varying as well.

In order to quantify the error made in approximating the nonlinear system with the LTV one we use the weighted L_∞ norm distance between the nonlinear and linear trajectories, that is, for each LTV system, and thus for each block Π_k of the partition, we define an approximation error function $e_k : \Pi_k \rightarrow \Re_+$ as

$$e_k(\pi) := \left\| y(t;\pi) - y_{Lk}(t;\pi) \right\|_\infty^b \tag{8}$$

We search for an approximation of the nonlinear system that introduces a pre-specified bounded error. Equivalently, this can be seen as searching for a partition $\{\Pi_k\}_L$ in which $e_k(\cdot)$ is below a pre-specified tolerance ε for all π in Π:

$$\{\Pi_k\}_L : \forall \Pi_k \in \{\Pi_k\}_L \quad \max_{\pi \in \Pi k} e_k(\pi) \le \varepsilon \tag{9}$$

As we will discuss later on, finding such a partition allows using the solution to problem 1 obtained for the LTV systems to approximate the one of the nonlinear system. Assumption 1 assures that a partition complying to Eq. (9) may always be found. Indeed, for any $\pi_k^0 \in \Pi_k$ we have:

$$\lim_{\pi \to \pi_k^0} e_k(\pi) = 0 \tag{10}$$

Thus, by using a partition of Π with sufficiently small blocks, it is possible to approximate as closely as desired the nonlinear trajectories using the LTV ones. Following this fact, an algorithm for finding $\{\Pi_k\}_L$ may be obtained by repeatedly shrinking the blocks of the partition for which the approximation error is higher than ε. The partition refinement is here obtained iteratively, by means of an isotropic bisection technique. The isotropic bisection procedure splits a single *p*-dimensional hyper-rectangle set in 2^p hyper-rectangular subsets, collectively exhaustive and mutually exclusive with respect to the "father" set. These "sons" are generated bisecting in each of the *p* dimensions the father hyper-rectangle's edges. At each iteration, the approximation error in each block Π_k is analyzed. Three cases are possible:

1. $\max_{\pi \in \Pi k} e_k(\pi) \le \varepsilon$. The error is below the tolerance. Π_k is assigned to $\{\Pi_k\}_L$: $\{\Pi_k\}_L = \{\Pi_k\}_L \cup \Pi_k$.
2. $\max_{\pi \in \Pi k} e_k(\pi) > \varepsilon$. The approximation error is higher than the tolerance. We shall split this condition into two further cases, depending on the volume of Π_k:
   a. The volume of Π_k is smaller than a predefined maximum resolution η, i.e. $\mathrm{vol}(\Pi_k) \le \eta$. In these blocks the system nonlinearities are so large as to prevent its LTV approximation within a small volume η and thus are not further considered for the subsequent step of the proposed algorithm. Such blocks are left undetermined from the gain tuning standpoint.
   b. The volume of Π_k is higher than η. Π_k is then partitioned into 2^p sons and the process of evaluating the maximum approximation error is repeated for each of them.

The major challenge in applying the above algorithm resides in the evaluation of the $e_k(\cdot)$ function's upper bound over a given Π_k, that is, in determining if the distance between the nonlinear and linear trajectories under the same π is within the tolerance for all $\pi \in \Pi_k$, as discussed in the next section.

3.1.1 Evaluation of nonlinear trajectories approximation error

A few approaches exist that allow relating the time responses of a nonlinear system to those of its linearization by quantitative means. These approaches conservatively bound from above a certain measure of the trajectories distance by maximizing some nonlinear time-varying test function over a vector space. They thus either solve an optimization problem, with related computational burden, or require prior knowledge of the test function maximum bound, for instance using the Lipschitz constant (Asarin et al., 2007) or the maximum bound of the dynamical function's second order derivatives (Desoer &

Vidyasagar, 1975). The latter methods, however, provide bounds on the trajectory distance that are typically exponentially increasing with time. This implies that in practice they can be used for time horizons of limited duration w.r.t. the system time-scales, which is not the case of re-entry applications. Alternative approaches have been proposed, which estimate the approximation error introducing some heuristic methods. In (Rewienski & White, 2001) the linear system is considered a valid approximation within a norm-ball, whose radius is determined depending on the linear trajectory characteristics. In (Tancredi et al., 2008) the approximation error over a polytope in the parameters space is estimated by its maximum value over the polytope's vertices, assuming that the polytope is sufficiently smaller than the scale at which the system exhibits significant nonlinear behaviour so that the maximum error always occurs in a vertex.

The approximation error is here evaluated in probabilistic terms, as proposed in (Tancredi et al., 2009). In particular, by fictitiously introducing a statistical description of the gains in the generic Π_k, we accept the risk of the approximation error being higher than the tolerance in a subset of Π_k having small probability measure. More precisely, we consider the nonlinear system to be well approximated in Π_k if the risk of $e_k(\cdot)$ being higher than the error tolerance is smaller than a threshold. The value of this threshold shall be selected sufficiently small as to avoid that $e_k(\cdot)$ can be higher than the tolerance with significant probability. However, it shall also be sufficiently high as to avoid that the Π_k sets have a volume smaller than the maximum resolution η. Preliminary numerical analyses suggest that in our problem setting a threshold value equal to 6% is a good compromise:

$$\Pi_k : Pr(e_k > \varepsilon) \leq 0.06 \Rightarrow \max_{\pi \in \Pi k} e_k(\pi) \leq \varepsilon \tag{11}$$

Without introducing any assumption on the probability distribution of $e_k(\cdot)$, we can then use the one-sided Chebyshev inequality to translate Eq.(11) in:

$$\Pi_k : E(e_k) + 4\sqrt{Var(e_k)} \leq \varepsilon \quad \Rightarrow \quad \max_{\pi \in \Pi k} e_k(\pi) \leq \varepsilon \tag{12}$$

In order to determine the mean and variance of $e_k(\cdot)$, we use the Scaled Unscented Transformation (SUT), first introduced in (Julier, 2002). More specifically, let us consider a generic hyper-rectangle $\Pi_k := [\underline{\pi}_{1k}, \overline{\pi}_{1k}] \times \ldots \times [\underline{\pi}_{pk}, \overline{\pi}_{pk}]$. We fictitiously assume π to be uniformly distributed in Π_k, which results in the following mean and covariance matrix, E_k and Cov_k, respectively.

$$E_k = \pi_k^0 \; ; \; Cov_k = \frac{1}{12} \times diag\left[\left(\overline{\pi}_{k1} - \underline{\pi}_{k1}\right)^2, \ldots, \left(\overline{\pi}_{kp} - \underline{\pi}_{kp}\right)^2 \right] \tag{13}$$

Using the SUT, we may estimate $e_k(\cdot)$ mean and variance. Specifically, according to (Julier, 2002), we choose a series of $2p+1$ points $\Theta_{ki} \in \Pi_k$, symmetrically distributed around the mean π_k^0, as follows:

$$\begin{aligned} &\Theta_{ki} = \pi_k^0 && i = 0 \\ &\Theta_{ki} = \pi_k^0 + \left[\sqrt{(p+\chi)Cov_k}\right]_i && i = 1, \ldots, p \\ &\Theta_{ki} = \pi_k^0 - \left[\sqrt{(p+\chi)Cov_k}\right]_{i-p} && i = p+1, \ldots, 2p \end{aligned} \tag{14}$$

where $\chi = \mu^2(p+\kappa) - p$ is a tunable parameter. Each point has an associated weight for computing the mean and variance of the $e_k(\cdot)$ function. The first point Θ_{k0} has instead two weights, one for computing the mean and one for the variance. Denoting with $W_i^{(m)}$ the weights for computation of E(e_k), and with $W_i^{(c)}$ the weights for computation of Var(e_k), the following hold (Julier, 2002):

$$W_0^{(m)} = \chi/(p+\chi) \; ; \; W_0^{(c)} = W_0^{(m)} + (1-\mu^2+\beta) \qquad (15a, b)$$

$$W_i^{(m)} = W_i^{(c)} = 1/[2(p+\chi)] \quad i=1,...,2p \qquad (15c)$$

The mean and variance of the $e_k(\cdot)$ function can be then estimated as the weighted average and weighted outer product of the transformed points, allowing to evaluate if Eq.(12) holds by numerically evaluating $2p + 1$ times the $e_k(\cdot)$ function :

$$E[e_k] = \sum_{i=0}^{2p} W_i^{(m)} e_k(\Theta_{ki}) \qquad (16a)$$

$$Var[e_k] = \sum_{i=0}^{2p} W_i^{(c)} \{E[e_k] - e_k(\Theta_{ki})\}^2 \qquad (16b)$$

The SUT has three tunable parameters, μ, β, and κ. Guidelines for tuning these parameters are given by (Van der Merwe et al., 2000), which suggests letting κ=0. β is a non-negative weighting term which can be used to incorporate knowledge of the higher order moments of the distribution. Preliminary numerical analyses have shown that in our problem setting β=0 delivers the best estimates. At last, μ controls the "size" of the Θ points distribution and should be $0 \le \mu \le 1$. We choose $\mu = (3/p)^{0.5}$ in order to have the Θ_{ki} points in the center of Π_k facets. This choice allows for sharing some computations between adjacent Π_k sets, and thus to reduce the overall computational load.

3.2 Property clearance

Once the $\{\Pi_k\}_L$ partition has been determined, one can obtain a solution to problem 1 by formulating a similar problem on the LTV approximating systems corresponding to $\{\Pi_k\}_L$. For such LTV systems, the difference between any nonlinear and linear trajectories under the same π is included in the closed ball in $\Re^n$ with respect to the norm in Eq. (8), with radius equal to ε, B_ε. It follows that the nonlinear solutions tube is included in the Minkowski sum between the solutions tube of its linearization and the former ball. To exploit this result in achieving the problem's solution, let us define a reduced admissible solution tube, obtained by shrinking $S_A(\cdot)$ of an amount equal to B_ε. Denoting as ⊕ the Minkowski sum operator, the reduced admissible solution tube reads:

$$S'_A(t): \; S'_A(t) \oplus B_\varepsilon = S_A(t) \;\; \forall t \in [0,T] \qquad (17)$$

The $S'_A(\cdot)$ complying to Eq. (17) can be easily determined since $S_A(\cdot)$ is hyper-rectangular at each time epoch. Being the norm sphere hyper-rectangular as well by definition, $S'_A(\cdot)$ can be obtained simply by component-wise difference of $S_A(\cdot)$ and B_ε. Consider now a modification of the P property, expressed in terms of $S'_A(\cdot)$ and of the linear trajectories corresponding to $\{\Pi_k\}_L$:

$$P'(\pi) := \begin{cases} true & y_{Lk}(t;\pi) \in S'_A(t) \ \forall t \in [0,T] \\ false & \exists t \in [0,T] : y_{Lk}(t;\pi) \notin S'_A(t) \end{cases} \tag{18}$$

It can be easily proved that P' implies P. Therefore, introducing a region of admissible uncertainties analogous to Π_A, but based on P', as $\Pi'_A := \{ \pi \in \Pi \mid P'(\pi) = true \}$, it follows that $\Pi'_A \subseteq \Pi_A$. It will be shown later on that a technique exists to obtain a good estimate of Π'_A. This is equivalent to obtain a conservative solution to problem 1, in the sense that the computed region Π'_A will be included in the actual Π_A. Nonetheless, the amount of conservativeness in estimating Π_A is bounded, and can be reduced as required by reducing the approximation error tolerance, at the price of a higher computational load.

3.2.1 Computation of Π'_A

The computation of Π'_A is obtained by exploiting the preservation of convexity in LTV trajectories under constant inputs and by applying a set inversion algorithm, SIVIA (Set Inverter Via Interval Analysis), originally developed in the framework of Interval Analysis. We briefly recall here the algorithm main features relevant to the present context, referring the interested reader to (Jaulin et al., 2001) and the references therein.

Given the definition of Π'_A and P', the determination of Π'_A may be seen as a set inversion problem, which is defined as follows. Let f be a possibly nonlinear function from $\Re^n$ to $\Re^m$, and let Y be a subset of $\Re^m$. Set inversion is the determination of the reciprocal image: $X=\{x \in \Re^n \mid f(x) \in Y\} = f^{-1}(Y)$, which in our case is Π'_A itself. The SIVIA algorithm allows to compute two sub-partitions of Π, that is, partitions of a subset of Π, that are an inner and outer enclosure of Π'_A, denoted as $\underline{\Pi}'_A$ and $\overline{\Pi}'_A$ respectively.

$$\underline{\Pi}'_A \subset \Pi'_A \subset \overline{\Pi}'_A \tag{19}$$

The algorithm is iterative, and is initially applied to the partition $\{\Pi_k\}_L$. In order to determine if a block Π_k belongs to the enclosures, it performs an *inclusion test* $[P']$, having the following properties:

$$[P'](\Pi_k) = true \quad \Rightarrow \quad \forall \pi \in \Pi_k, P'(\pi) = true \tag{20a}$$

$$[P'](\Pi_k) = false \quad \Rightarrow \quad \forall \pi \in \Pi_k, P'(\pi) = false \tag{20b}$$

More precisely, the inner enclosure $\underline{\Pi}'_A$ is composed of hyper-rectangular blocks Π_k for which the inclusion test is true. Given Eq. (19), such blocks are also members of $\overline{\Pi}'_A$. Reversely, if it can be proved that $[P'](\Pi_k) = false$, then the block has an empty intersection with Π'_A, and it is thus rejected. Otherwise, no conclusion can be drawn based on the inclusion test, and the block Π_k is called undetermined. The latter is then bisected in 2^p subsets that are tested until their volume reaches the user-specified resolution η. Thus, such undetermined minimum-volume blocks are deemed small enough to be stored in the outer approximation $\overline{\Pi}'_A$ of Π'_A.

3.2.2 Inclusion test for SIVIA

The application of SIVIA requires defining an inclusion test, which is typically obtained by applying interval analysis, e.g. in (Juliana et al., 2008). However, interval computation is usually pessimistic, in the sense that a block Π_k may be deemed undetermined by an inclusion test even if the property under analysis holds uniformly (i.e. attains the same Boolean value) over the block itself. This implies a substantial increase in the computational load, which is particularly critical since the algorithm computational complexity increases exponentially with p. In the present context, we use the inclusion test proposed in (Tancredi et al., 2009), which captures exactly the blocks in which P' is uniformly true, and also provide a condition which is sufficient for P' to be uniformly false.

The inclusion test exploits the preservation of convexity in LTV trajectories under constant inputs. Consider a generic hyper-rectangular $\Pi_k \in \{\Pi_k\}_L$. Π_k is a convex polytope having 2^p vertices $\pi_k^{(v)}$, i.e. it admits the following vertex representation:

$$\Pi_k = \left\{ \pi \in \Re^p \middle| \pi = \sum_{v=1}^{2^p} \lambda_{kv} \pi_k^{(v)},\ \lambda_{kv} \geq 0,\ \sum_{v=1}^{2^p} \lambda_{kv} = 1 \right\} \tag{21}$$

Because the trajectory of an LTV system under a constant input π may be viewed as an affine transformation with respect to π, any solution of the LTV system under a generic π in Π_k is a convex combination of the solutions under all the $\pi_k^{(v)}$. The output trajectories $y_{Lk}(t;\pi)$ thus span the following set (tube) $S_{Lk}(t)$ as π varies in Π_k.

$$S_{Lk}(t) = \left\{ y_{Lk}(t;\pi) \in \Re^w \middle| y_{Lk}(t;\pi) = \sum_{v=1}^{2^p} \lambda_{kv} y_{Lk}\left(t;\pi_k^{(v)}\right), \lambda_{kv} \geq 0,\ \sum_{v=1}^{2^p} \lambda_{kv} = 1 \right\} \tag{22}$$

As a consequence, the knowledge of the 2^p vertex trajectories $y_{Lk}(t;\pi_k^{(v)})$ allows to determine exactly the solutions tube corresponding to Π_k. We exploit this property to define the inclusion test, which is a comparison of the two time varying polytopes S_{Lk} and S'_A. More precisely, the condition $S_{Lk}(t) \subseteq S'_A(t)$ for all $t \in [0,T]$ is equivalent to P' being uniformly true in Π_k, and, given S_{Lk} convexity, it is equivalent also to the 2^p vertex trajectories $y_{Lk}(t;\pi_k^{(v)})$ belonging to $S'_A(t)$. The condition $S_{Lk}(t) \cap S'_A(t) = \emptyset$ for at least one $t \in [0,T]$ is instead equivalent to P' being uniformly false in Π_k. Unfortunately, this condition may not be checked using only the knowledge of the vertex trajectories, but would require further computations to be ascertained exactly. We instead provide a condition involving only the vertex trajectories, which is only sufficient for $S_{Lk}(t) \cap S'_A(t) = \emptyset$. In particular, we exploit the fact that $S'_A(\cdot)$ is hyper-rectangular by assumption, and thus admits an easily-obtainable half-space representation $S'_A(t) = \{y \in \Re^w \mid S_A^L y \leq S_A^R(t)\}$, where $S_A^L = (I_w, -I_w)^T$, $S_A^R:[0,T] \rightarrow \Re^{2w \times 1}$ (I_w stand for the w by w identity matrix). In case at least one of the $2w$ inequalities defining $S'_A(\cdot)$ is not satisfied by all the vertex trajectories, the solutions tube $S_{Lk}(\cdot)$ lies completely outside $S'_A(\cdot)$, implying $S_{Lk}(t) \cap S'_A(t) = \emptyset$. We thus define the following inclusion test, which formally resumes the previous discussion. Its evaluation requires a limited (and known a priori) number of linear trajectories, which are obtained by numeric simulation. Note that nonlinear simulations are not needed for the evaluation of the inclusion test, which involves only simulation of the linear approximations.

$$[P'](\Pi_k) := true \Leftrightarrow \quad \begin{array}{l} \forall t \in [0,T], \forall \nu = 1,..,2^p \\ S_A^L y_{Lk}\left(t;\pi_k^{(\nu)}\right) \le S_A^R(t) \end{array} \tag{23a}$$

$$[P'](\Pi_k) := false \Leftarrow \quad \begin{array}{l} \exists t \in [0,T], \exists i = 1,..,2m: \forall \nu = 1,..,2^p \\ \left[S_A^L\right]_i y_{Lk}\left(t;\pi_k^{(\nu)}\right) > \left[S_A^R(t)\right]_i \end{array} \tag{23b}$$

Applying the above procedure, Π'_A is determined exactly within a prefixed resolution, and, due to the properties of the LTV systems defined on $\{\Pi_k\}_L$, problem 1 is solved conservatively for the nonlinear system (1).

4. Application case

This section introduces the nonlinear system describing the closed-loop longitudinal flight dynamics of the experimental reusable launch vehicle demonstrator USV-FTB1 currently operated by the CIRA. The FTB1 vehicle is the first of three planned vehicle configurations that CIRA is developing as part of its USV Program, whose main goal is contributing to the international community effort toward the development of next generation reusable space vehicles. This vehicle is planned to execute flight tests in subsonic, transonic and low supersonic flight regimes, in view of the development of upgraded vehicle configurations to perform sub-orbital and orbital re-entry flights. The FTB1 vehicle, described in detail in (Russo, 2009) and shown in Fig. 1, is unmanned and un-powered. It has a slender wing configuration, with two sets of aerodynamic effectors: the elevons, which provide pitch control when deflected symmetrically and roll control when deflected asymmetrically, and the rudders for yaw control.

Fig. 1. USV-FTB1 vehicle.

The second Dropped Transonic Flight Test (DTFT-2) of the FTB1 vehicle is specifically considered to show the effectiveness of the proposed technique. The mission, successfully executed in April 2010, foresees a drop of the vehicle from a stratospheric balloon (at nearly null velocity and angle of attack) to reach Mach numbers in the range of [1.2÷1.3] for investigating aerodynamics and advanced guidance navigation and control in the transonic phase of an un-powered re-entry flight. The basic operations of the mission consist of an ascent phase during which the stratospheric balloon brings the FTB1 at the release altitude of about at about 24-26 km followed by a flight phase where the FTB1 is dropped and the aerodynamic controlled flight starts. The vehicle accelerates until the desired Mach number is reached, and then starts a Mach-hold phase in which it performs a sweep in angle of attack for maintaining a constant Mach number. A deceleration phase is then initiated (up to 0.2 Mach) at the end of which a recovery parachute is deployed. The mission ends with the demonstrator splash down in the Mediterranean Sea.

Because the scope of the present section is to demonstrate the effectiveness of the gain tuning technique on an application of practical engineering relevance, we will restrict the analysis to a simplified version of the longitudinal FCL of the FTB1 vehicle, which was used in the initial design phases for executing flight mechanics analyses. Note that the FCL analyzed in this section is significantly different from the ones implemented for the DTFT2 mission (see, for instance, Morani et al., 2011, for a detailed description of the guidance law). Given the nonlinear augmented longitudinal dynamics of the FTB1 vehicle in the DTFT2 mission, the aim of the present analysis is to find (a set of) the controller's gains compliant to a requirement expressed as inclusion in a solution's tube.

The FTB1 vehicle longitudinal dynamics are modelled by means of standard nonlinear equations (Etkin & Reid, 1996), yielding a sixth order model. Actuator dynamics are included by means of a second order system and first order filters are used for modelling the navigation sensors for a and q. The longitudinal dynamics are augmented by a proportional-derivative flight control law, arranged in a cascade structure with feedback on the pitch rate q and angle of attack a. The augmented vehicle is driven by a time-varying angle of attack reference signal a_{ref}, which ramps up from zero at the vehicle release from the stratospheric balloon up to 8 deg. in the initial drop phase. The angle of attack is held constant until the desired Mach number of about 1.2 is reached. The Mach hold phase follows, where an α - sweep manoeuvre is performed. At the end of the Mach hold phase, the angle of attack increases up to 10 deg., value maintained in low subsonic conditions until parachute deployment. The overall feedback action is shown in Fig. 2 and has the following analytical expression, where ζ stands for the FCL internal state.

$$\delta_e = k_3\left[k_1\left(\alpha_{ref} - \alpha\right) + \zeta - q\right]; \quad \dot{\zeta} = k_1 k_2\left(\alpha_{ref} - \alpha\right) \tag{24}$$

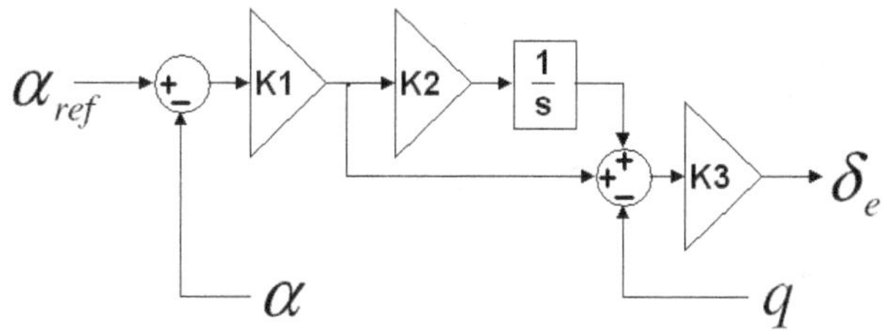

Fig. 2. FCL structure.

Two of the three gains of the controller, k_1, k_2, and k_3 are scheduled depending on the dynamic pressure q_∞, according to $k_1 = k_{10}$, $k_2 = k_{20} + k_{2s}\, q_\infty$, and $k_3 = k_{30} + k_{3s}/q_\infty$. The FCL foresees thus five gains. The nominal gains setting is $k'_1 = 1.25\ s^{-1}$; $k'_{20} = -1.28\ s^{-1}$;

$k'_{2s} = 4.48 \cdot 10^{-4}$ m · s ·kg $^{-1}$; $k'_{30} = 1.73 \cdot 10^{-1}$ s ; $k'_{3s} = -7.76 \cdot 10^{3}$ kg ·s $^{-1}$ ·m $^{-1}$. These gains have been determined applying standard LTI control synthesis techniques, and the resulting control law yields satisfactory LTI stability characteristics. Because of the complexity of the LTI based analysis when applied to these vehicles flying markedly time-varying trajectories, and because the focus of the present chapter is on determining the effectiveness of the proposed technique in dealing with time-based control performance requirements, the results of the stability analysis are not shown here for brevity. The reader is referred to (Tancredi et al., 2011) for an overview of the LTI stability analysis in a similar application.

The nominal response is obtained applying the above gain tuning, and considering the system to start at t_0 = 21.55 s. This is the first time epoch at which the Mach number is at least equal to 0.7, i.e. $M \geq 0.7$, which is the threshold condition above which the actuation system gains sufficient command authority for controlling the angle of attack.

The nominal response's angle of attack and commanded elevon deflections δ_e are shown in Fig. 3. The initial oscillation in a is caused by a sharp decrease of the elevons efficiency in the transonic phase. However, because of the considerable uncertainty on the entity of this phenomenon, no dedicated feed-forward actions were implemented.

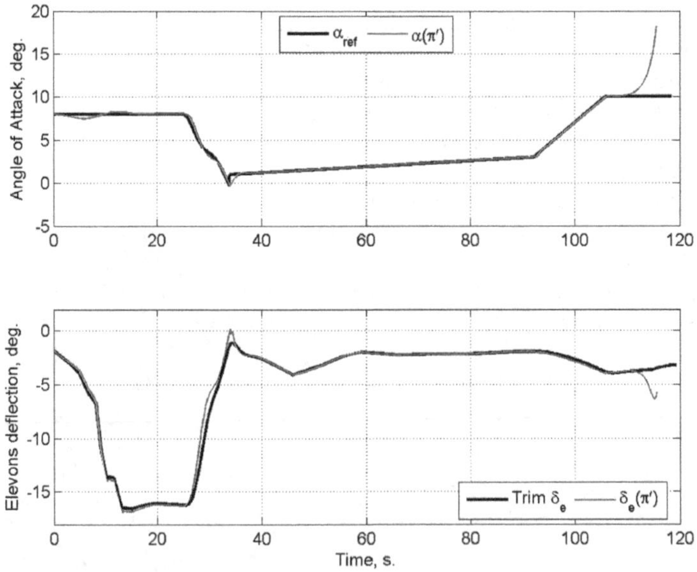

Fig. 3. Nominal response time histories.

The tuning analysis has been performed on a subset of the FCL gains. More precisely, we analyzed effects on the variations of the three most influential gains: the proportional α constant gain k_1, the scheduling gain of the integral α action k_{2s} and the scheduling gain of the proportional q action k_{3s}. In other words, we analyze the effects of the three dimensional vector $\pi := (k_1 \;\; k_{2s} \;\; k_{3s})^T$, in the following hyper-rectangular domain K, centred about the nominal tuning π'.

$$\Pi := [0,\ 2k'_1] \times [0,\ 2k'_{2s}] \times [2k'_{3s},\ 0]$$

The maximum allowed distances of the above variables for a meaningful linearization are set to 0.2 deg in angle of attack, 1.8 deg s^{-1} in pitch rate and 1.5 deg in elevons deflection. The admissible solutions tube constrains only the angle of attack and the elevons deflections. Elevons deflection are required to be within [-20, 20] deg., which represent the limits of the actuation system. The solutions tube in α is tailored around the reference signal a_{ref}, enforcing the required maximum tracking error of ±0.6 deg. Because of the previously mentioned oscillation, the tracking requirement is relaxed to ±0.8 deg. in the transonic phase. The final α hold phase is treated separately from the remainder of the trajectory. Indeed, both tracking requirements are less stringent in this phase, increasing up to ± 2 deg., and the vehicle flight performances are dramatically different in these low subsonic flight conditions than in the remainder of the trajectory. Separating the tracking requirements in these two parts of the trajectory allows for a clearer understanding of the method potentials. Because of this setting, two admissible solution tubes are introduced: the final tube, which enforces requirements only on the final α-hold phase, and the tracking tube, which enforces tracking requirements in the remainder of the trajectory. Note that since the linearization error is taken into account in the admissible solutions tube definition (see section 3.2), the a_{ref} tracking requirements to which the linearized solutions shall comply are tighter than the enforced ones of ± 0.2 deg. Fig. 4 shows the two required solutions tubes in α, as well as the above mentioned "reduced" bounds. Note that the nominal tuning does not comply with any of the two tubes.

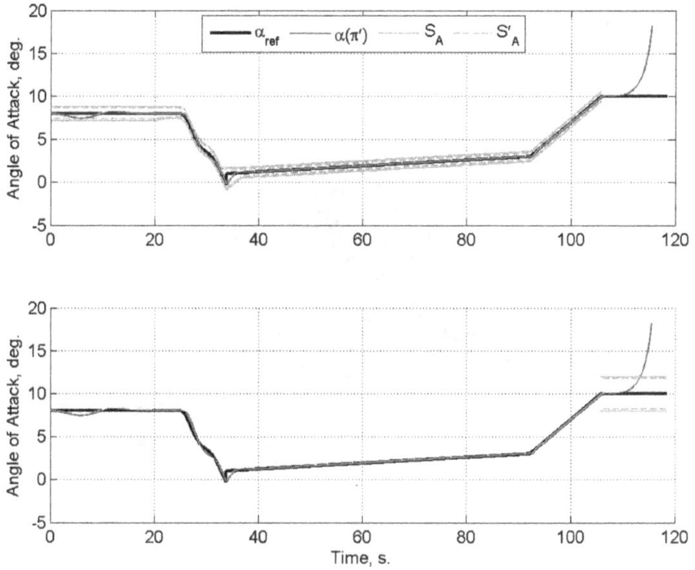

Fig. 4. Admissible solution tubes. Tracking (top) and Final (bottom).

4.1 Results

The approximation phase results are collected in Fig. 5, showing $\{\Pi_k\}_L$, the partition into which the gain domain Π has been divided to obtain a meaningful linearization. Results show that the original nonlinear system is successfully approximated only in a subset of the gain domain. In the remainder of Π, the system state vector dependency on the gains is highly nonlinear, and prevents the system to be approximated by its time-varying linearization even in Π_k subsets with the minimum allowed volume η (see section 3.1). The approximation phase results were obtained with a CPU time of ~ 10 hours on a standard personal computer. Note however, that its results do not depend on the admissible solutions tube, and are thus used for both the tracking and the final ones without the need of computing the approximation twice. The property clearance phase calls for a computational load that is only a fraction of the approximation one. In fact, evaluation of the inclusion test in Eq.(23) needs the numerical simulation only of the linear approximating systems, and nonlinear simulations are not involved at all. Fig. 6 collects the clearance phase results for the tracking tube. It can be seen how the inclusion test of Eq.(23) divides the blocks of the partition $\{\Pi_k\}_L$. The property clearance phase builds upon simulation of the linear approximations, and thus requires only about 1 hour of computation time. The region in which the gains comply with the tracking requirements, Π'_A, is shown in Fig. 7 for both the tracking and final tubes. As anticipated, the compliance region of both tubes does not comprise the nominal tuning. Compliance to each of the two tubes calls for lower than nominal values of the scheduling gain of the proportional q action, k_{3s} , coupled to higher proportional and integral scheduling gains of α, k_1 and k_{2s} , respectively. Requirements yielding to the final tube, however, are much more restrictive than those in the remainder of the trajectory, as can be seen by the small dimensions of the corresponding Π'_A region.

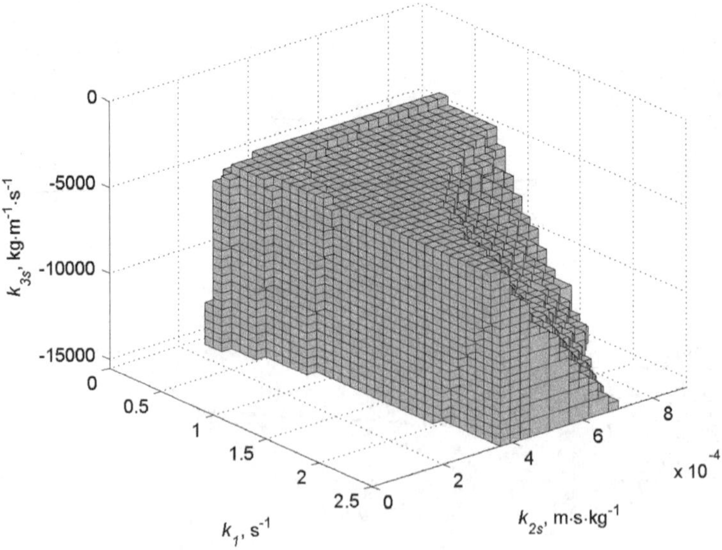

Fig. 5. Approximation results: $\{\Pi_k\}_L$

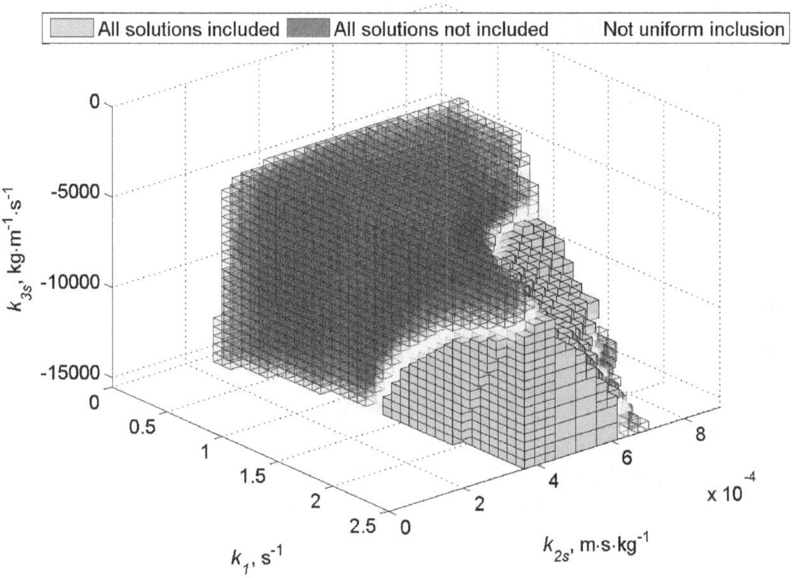

Fig. 6. Property Clearance results - Tracking tube.

These results demonstrate one of the main advantages of the proposed approach, that is, the capability to support the physical understanding of all the causes for unsatisfactory performances of the FCL within the whole Π region, being confident of having covered all possible gain combinations of interest. Fig. 8 compares the compliance regions of the two tubes, which are disjoint by a very small offset. However, because the offset dimensions are comparable to the resolution at which the results have been obtained, the true compliance region of the tracking tube may extend as to intersect the final tube's one. Even if this may in principle also not be the case, common sense suggests that a tuning lying near this offset would have tracking performances that do not violate significantly both tubes.

At last, we present the nonlinear system's simulation for a candidate tuning. In order to select this "optimal" tuning, π_{opt}, we choose the root mean square (RMS) of the a_{ref} tracking error as a cost function. The so obtained optimal tuning is shown in Fig. 8 as well, and is compared to the nominal one in Table 1 and in Fig. 9. Results show that the optimal gain yields a significantly smaller RMS error in tracking α_{ref} than the nominal one, and improves the system behaviour in the final phase.

Gains	**k_1, s^{-1}**	**; k_{2s}, m·s·kg^{-1}**	**; k_{3s}, kg·s^{-1}·m^{-1}.**	**RMS error, deg**
π'	**1.25**	**4.48·10^{-4}**	**-7.76·10^{3}**	**0.34**
π_{opt}	**1.81**	**3.21·10^{-4}**	**-8.54·10^{3}**	**0.15**

Table 1. Comparison of nominal and optimal gains.

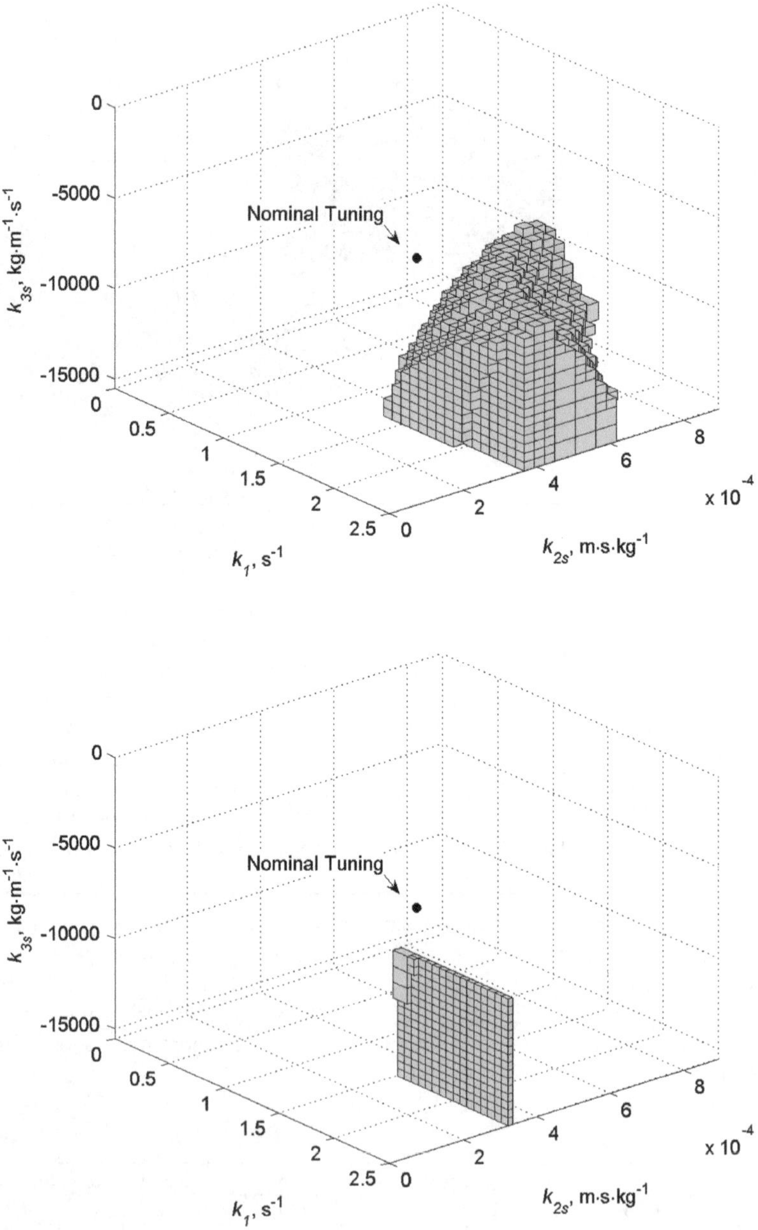

Fig. 7. Nominal tuning vs. compliance region Π'_A. Tracking tube (top) and Final Tube (bottom)

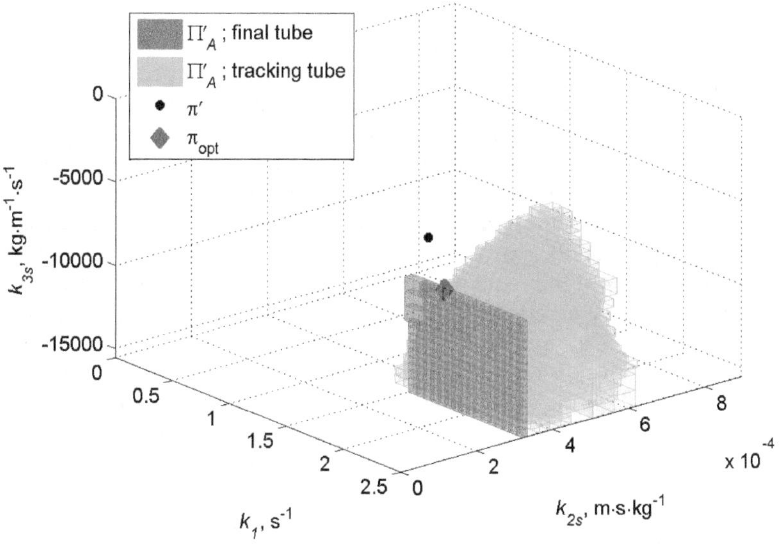

Fig. 8. Comparison between tracking and final tubes compliance regions.

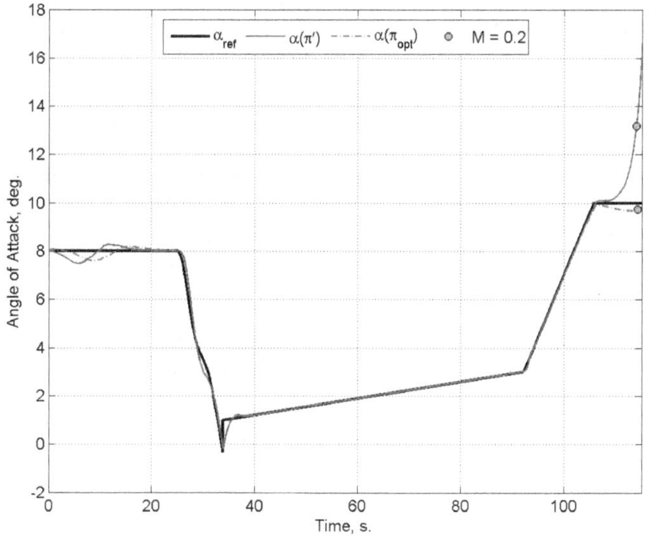

Fig. 9. Selected gains nonlinear simulations: α time history.

5. Conclusion

A novel approach to gain tuning has been developed, based on previous results that were obtained by the authors for a different problem. The approach specifically applies to gliding vehicles in the terminal phases of re-entry flight, and is capable of handling gain scheduled control laws under trajectory tracking requirements. Its capability of highlighting the causes for requirement violation, being confident of having covered all possible combinations of the controller gains, makes the developed technique an effective tool for driving the control law refinement, as shown in an application of practical engineering significance. The adoption of practical stability as a criterion for enforcing trajectory tracking requirements is promising thanks to its inherent capability of handling the original mission or system requirements. In fact, it allows taking explicitly into account trajectory time-varying effects in the tuning task, which can be significant for the applications of interest. The practical stability approach improves the accuracy in evaluating the control law performances with respect to frozen-time approaches, thus reducing the risk of highlighting effects that were not previously disclosed when applying numerical verification methods, such as Monte Carlo techniques. This would avoid the need of upgrading a control design tuning with scarce information on the causes for unsatisfactory performance, as it typically occurs when applying numerical verification methods early in the design cycle, thus streamlining the overall design cycle. In this sense, the proposed approach is though to be complementary both to classical LTI-based design tools and to numerical verification methods.

One important issue of the method is in the number of gains that can be simultaneously treated, due to the exponential increase in the computational load. Nonetheless, its application so far suggests that, when the method is executed on a standard desktop computer, the maximum dimension of manageable problems is in the order of five, depending on the features of the specific application case, most notably its nonlinearity in the whole uncertainty domain. For the application shown in the chapter, the map relating the system state vector to gain values was determined to be heavily nonlinear. This feature is thought to be distinctive of most gain tuning problems, as suggested by common sense and relevant literature, even though further investigations would be needed for ascertaining this claim. This pronounced nonlinearity further limits the method applicability because accurate linear approximations are valid only in small subsets of the gain domain, thus calling for a refined partition, which causes an increase in the computational load. Nonetheless, distributed computing and the use of more powerful computing machines substantially increase the number of gains that can be taken into account.

At last, the presented approach is based on the practical stability criterion, which allows translating tracking requirements in terms of the maximum tracking error. However, in most trajectory tracking applications, the RMS tracking error is also included in the requirements. Note that the RMS error is a convex function of the tracked variable. As such, defining an opportune Boolean property being true when the RMS error is below a certain threshold, one should be capable of devising an inclusion test similar to the one presented in this chapter. This would allow extending the approach for being capable of handling requirements on both maximum and RMS tracking errors. Further work will concern this possibility.

6. References

Amato F., Ariola, M., Cosentino, C., Abdallah, C.T., and Dorato, P. (2003). Necessary and Sufficient Conditions for Finite-Time Stability of Linear Systems, *Proceedings of the 2003 American Control Conference,* Vol. 5, pp. 4452- 4456.

Asarin, E., Dang, T., and Girard, A. (2007). Hybridization methods for the analysis of nonlinear systems, *Acta Informatica,* Vol.43, no. 7, pp. 451-476.

Crespo L. G., Kenny S. P., Giesy D. P. (2008). A Verification-driven Approach to Control Analysis and Tuning, *Proceedings of the AIAA Guidance Navigation and Control Conference,* Honolulu, Hawaii, AIAA-2008-6340.

Crespo, L. G., Matsutani, M., Annaswamy A. M. (2010). Verification and Tuning of an Adaptive Controller for an Unmanned Air Vehicle, *Proceedings of the AIAA Guidance Navigation and Control Conference.* Toronto, Ontario, Canada, AIAA-2010-8403.

Desoer, C. A., and Vidyasagar, M. (1975). *Feedback systems: input-output properties,* Academic Press, Inc., New York, Chap. 4.

Dorato, P. (2006). An Overview of Finite-Time Stability, *Current Trends in Nonlinear Systems and Control: In Honor of Petar Kokotovic and Turi Nicosia, ,* Birkhauser Boston, pp 185-195.

dos Santos Coelho, L. (2009). Tuning of PID controller for an automatic regulator voltage system using chaotic optimization approach, *Chaos, Solitons & Fractals,* Volume 39, Issue 4, Pages 1504-1514, ISSN 0960-0779.

Etkin, B., and Reid, L.D. (1996). *Dynamics of Flight: Stability and Control,* John Wiley and Sons, Inc, New York, 3rd ed.

Gruyitch, L., Richard, J-P., Borne P., and Gentina, J.C. (2000). *Stability Domains,* Chapman & Hall/CRC, Boca Raton, FL, Chaps. 1,2, 6.

Jaulin, L., Kieffer, M., Didrit, O., Walter, E. (2001). *Applied Interval Analysis, with Examples in Parameter and State Estimation, Robust Control and Robotics,* Springer-Verlag, London, Chaps. 2, 3.

Juliana, S., Chu, Q.P., Mulder, J.A. (2008). Reentry Flight Clearance Using Interval Analysis, *AIAA Journal of Guidance, Control, and Dynamics,* Vol.31, n.5, pp.1295-1307.

Julier, S. J. (2002). The scaled unscented transformation, *Proceedings of the 2002 American Control Conference,* Vol.6, pp. 4555–4559.

Lee, H. C., and Choi, J. W. (2004). Linear Time-Varying Eigenstructure Assignment with Flight Control Application. *IEEE Transactions on Aerospace and Electronic Systems,* Vol. 40, No. 1, pp. 145- 157.

Leith, D. J., and Leithead, W. E. (2000). Survey of gain-scheduling analysis and design, *International Journal of Control,* Vol. 73, No. 11, pp. 1001-1025.

Miyazawa Y., Motoda, T. (2001). Stochastic Parameter Tuning Applied to Space Vehicle Flight Control Design, *AIAA Journal of Guidance, Control and Dynamics,* Vol. 24, No. 3, pp. 597-604.

Morani, G, Cuciniello, G, Corraro, F, and Di Vito, V. (2011). On-line guidance with trajectory constraints for terminal area energy management of re-entry vehicles, *Proceedings of the Institution of Mechanical Engineers, Part G: Journal of Aerospace Engineering,* 225: 631.

Rewienski, M., White, J. (2001). A Trajectory Piecewise - Linear Approach to Model Order Reduction and Fast Simulation of Nonlinear Circuits and Micromachined Devices, *Proceedings of the 2001 IEEE/ACM international conference on Computer-aided design,* pp. 252 - 257.

Russo, G. (2009). DTFT-1: Analysis of the first USV flight test, *Acta Astronautica*, Volume 65, Issues 9-10, Pages 1196-1207, ISSN 0094-5765.

Ryali, V., Moudgalya, K.M. (2005). Practical stability analysis of uncertain nonlinear systems, *Proceedings of the National Conference on Control and Dynamic Systems*, I.I.T. Bombay.

Tancredi, U., Grassi, M., Corraro, F., Filippone, E., and Russo, M. (2008). A Novel Approach to Clearance of Flight Control Laws over Time Varying Trajectories", *Automatic Control in Aerospace*, Vol. 1, No. 1, Paper 2, Retrieved from: <http://www.aerospace.unibo.it/index.php?e=5>

Tancredi, U., Grassi, M., Corraro, F., and Filippone E. (2009). Robustness Analysis for Terminal Phases of Reentry Flight, *AIAA Journal of Guidance, Control, and Dynamics*, Vol. 32, No. 5, pp. 1679 - 1683.

Tancredi, U., Grassi, M., Corraro, F., Vitale, A., & Filippone, E. (2011). A linear time varying approach for robustness analyses of a re-entry flight technology demonstrator, *Proceedings of the Institution of Mechanical Engineers, Part G: Journal of Aerospace Engineering* , in press.

Van der Merwe, R., De Freitas, N., Doucet, A., and Wan, E. A. (2000). The Unscented Particle Filter, *Advances in Neural Information Processing Systems*, pp.584–590.

Wang, Q., and Stengel, R. F. (2002). Robust control of nonlinear systems with parametric uncertainty, *Automatica*, Vol. 38, No. 9, pp. 1591-1599.

3

Quantitative Feedback Theory and Its Application in UAV's Flight Control

Xiaojun Xing and Dongli Yuan
Northwestern Polytechnical University, Xi'an, China

1. Introduction

Quantitative feedback theory (hereafter referred as QFT), developed by Isaac Horowitz (Horowitz, 1963; Horowitz and Sidi, 1972), is a frequency domain technique utilizing the Nichols chart in order to achieve a desired robust design over a specified region of plant uncertainty. Desired time-domain responses are transformed into frequency domain tolerances, which lead to bounds (or constraints) on the loop transmission function. The design process is highly transparent, allowing a designer to see what trade-offs are necessary to achieve a desired performance level.

QFT is also a unified theory that emphasizes the use of feedback for achieving the desired system performance tolerances despite plant uncertainty and plant disturbances. QFT quantitatively formulates these two factors in the form of (a) the set $\Im_R = \{T_R\}$ of acceptable command or tracking input-output relationships and the set $\Im_D = \{T_D\}$ of acceptable disturbance input-output relationships, and (b) a set $\varphi = \{P\}$ of possible plants which include the uncertainties. The objective is to guarantee that the control ratio $T_R = Y / R$ is a member of $\Im_R$ and $T_D = Y / D$ is a member of $\Im_D$, for all plants P which are contained in φ. QFT has been developed for control systems which are both linear and nonlinear, time-invariant and time-varying, continuous and sampled-data, uncertain multiple-input single-output (MISO) and multiple-input multiple-output (MIMO) plants, and for both output and internal variable feedback.

The QFT synthesis technique for highly uncertain linear time-invariant MIMO plants has the following features:

1. The MIMO synthesis problem is converted into a number of single-loop feedback problems in which parameter uncertainty, external disturbances, and performance tolerances are derived from the original MIMO problem. The solutions to these single-loop problems represent a solution to the MIMO plant.
2. The design is tuned to the extent of the uncertainty and the performance tolerances.

This design technique is applicable to the following problem classes:

1. Single-input single-output (SISO) linear-time-invariant (LTI) systems
2. SISO nonlinear systems.
3. MIMO LTI systems.
4. MIMO nonlinear systems.

5. Distributed systems.
6. Sampled-data systems as well as continuous systems for all of the preceding.

Problem classes 3 and 4 are converted into equivalent sets of MISO systems to which the QFT design technique is applied. The objective is to solve the MISO problems, i.e., to find compensation functions which guarantee that the performance tolerances for each MISO problem are satisfied for all P in φ.

This chapter is essentially divided into two parts. The first part, consisting of Sections 2 through 4, presents the fundamentals of the QFT robust control system design technique for the tracking and regulator control problems. The second part consists of Seciton 5 which focuses on the application of QFT techinique to the flight control design for a certain Unmaned Aerial Vehicle (UAV). This is accomplished by decomposing the UAV's MIMO plant to 2 MISO plants whose controllers are both synthisized using QFT techique for MISO systems. And the effectiveness of both controllers is verified according the digital simulation results. Besides, Sections 6 through 8 are about summary of whole chapter, references and symbols used in the chapter.

2. Overview of QFT

2.1 Design objective of QFT

Objective of QFT is to design and implement robust control for a system with structured parametric uncertainty that satisfies the desired performance specifications.

2.2 Performance specifications for control system

In many control systems the output $y(t)$ must lie between specified upper and lower bounds, $y(t)_U$ and $y(t)_L$, respectively, as shown in Fig.1a. The conventional time-domain figures of merit, based upon a step input signal $r(t)$ are shown in Fig.1a. They are: M_p, peak overshoot; t_r, rise time; t_p, peak time; and t_s, settling time. Corresponding system performance specifications in the frequency domain are, B_U and B_L, the upper and lower bounds respectively, peak overshoot $Lm\,M_m$, and the frequency bandwidth ω_h which are shown in Fig.1b.

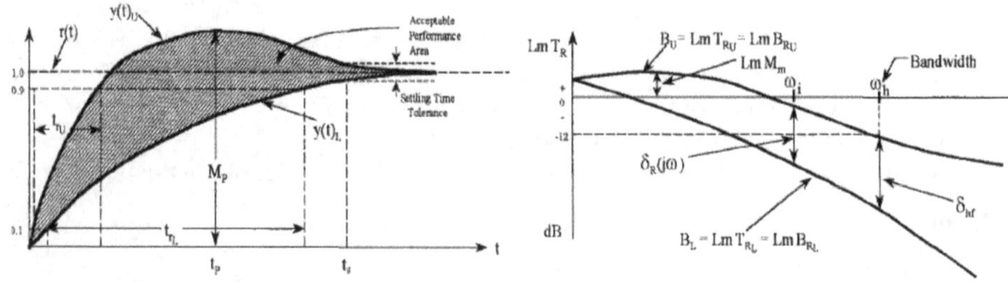

(a) time domain response specifications (b) frequency domain response specifications

Fig. 1. Desired system performance specifications

Assume that the control system has negligible sensor noise and sufficient control effort authority, then for a stable LTI minimum-phase plant, a LTI compensator may be designed to achieve the desired control system performance specifications.

2.3 Implementation of QFT design objective

The QFT design objective is achieved by:

- Representing the characteristics of the plant and the desired system performance specifications in the frequency domain.
- Using these representations to design a compensator (controller).
- Representing the nonlinear plant characteristics by a set of LTI transfer functions that cover the range of structured parametric uncertainty.
- Representing the system performance specifications (see Fig.1) by LTI transfer functions that form the upper B_U and lower B_L boundaries for the design.
- Reducing the effect of parameter uncertainty by shaping the open-loop frequency responses so that the Bode plots of the J closed-loop systems fall between the boundaries B_U and B_L, while simultaneously satisfying all performance specifications.
- Obtaining the stability, tracking, disturbance, and cross-coupling (for MIMO systems) boundaries on the Nichols chart in order to satisfy the performance specifications.

2.4 QFT basics

Consider the control system of Fig.2, where $G(s)$ is a compensator, $F(s)$ is a prefilter, and $\wp$ is the nonlinear plant with structured parametric uncertainty. To carry out a QFT design:

- The nonlinear plant is described by a set of J minimum-phase LTI plants, i.e., $\wp = \{P_t(s)\}(t = 1,2,\cdots,J)$ which define the structured plant parameter uncertainty.
- The magnitude variation due to the plant parameter uncertainty, $\delta_P(j\omega_i)$, is depicted by the Bode plots of the LTI plants as shown in Fig. 3 which is for a certain plant.
- J data points (log magnitude and phase angle), for each value of frequency, $\omega = \omega_i$, are plotted on the Nichols chart. A contour is drawn through the data points that described the boundary of the region that contains all J points. This contour is referred to as a template. It represents the region of structured plant parametric uncertainty on the Nichols chart and are obtained for specified values of frequency, $\omega = \omega_i$, within the bandwidth (BW) of concern. Six data points (log magnitude and phase angle) for each value of ω_i are obtained, as shown in Fig. 4a, for a certain example to plot the templates, for each value of ω_i, as shown in Fig. 4b.
- The system performance specifications are represented by LTI transfer functions, and their corresponding Bode plots are shown in Fig. 3 by the upper and lower bounds B_U and B_L, respectively.

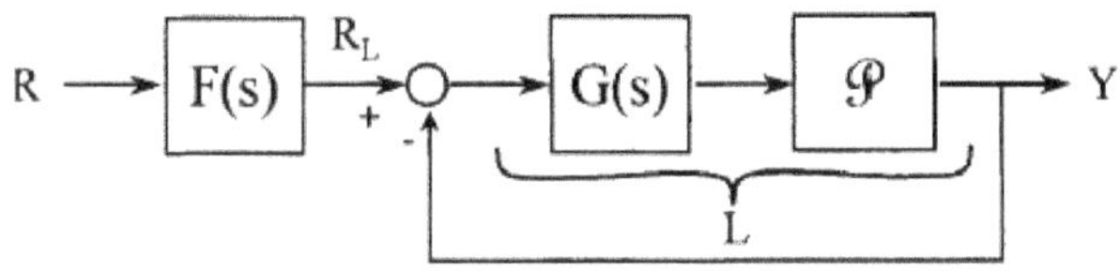

Fig. 2. Compensated nonlinear system

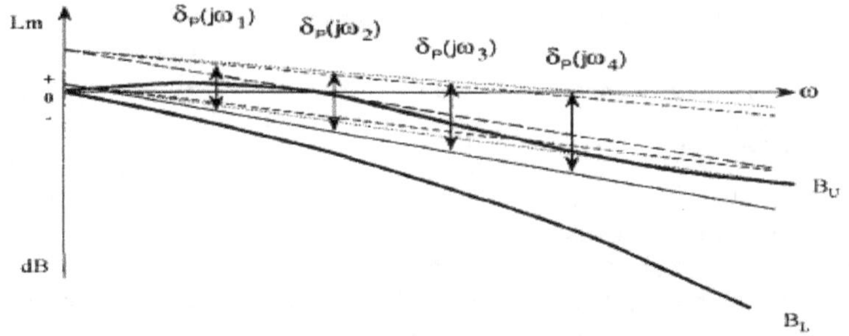

Fig. 3. LTI plants

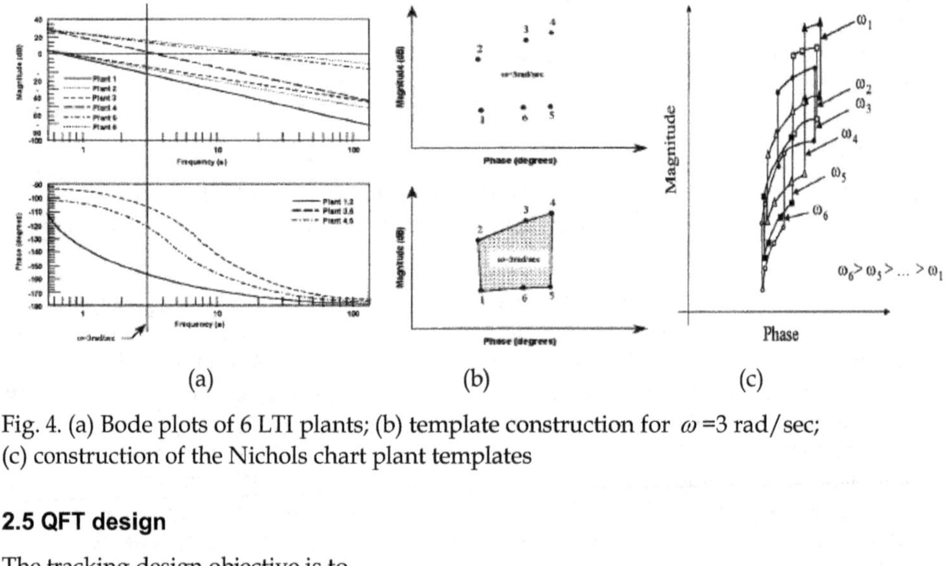

(a) (b) (c)

Fig. 4. (a) Bode plots of 6 LTI plants; (b) template construction for ω =3 rad/sec; (c) construction of the Nichols chart plant templates

2.5 QFT design

The tracking design objective is to

a. Synthesize a compensator $G(s)$ of Fig. 2 that
   - results in satisfying the desired performance specifications of Fig. 1
   - results in the closed-loop frequency responses T_{Li} shown in Fig. 5
   - results in the $\delta_L(j\omega_i)$ of Fig. 5 of the compensated system, being equal to or smaller than $\delta_P(j\omega_i)$ of Fig. 3 for the uncompensated system and that it is equal or less than $\delta_R(j\omega_i)$, for each value of ω_i of interest; that is: $\delta_L(j\omega_i) \le \delta_R(j\omega_i) \le \delta_P(j\omega_i)$

b. Synthesize a prefilter $F(s)$ of Fig. 2 that results in shifting and reshaping the T_{Li} responses in order that they lie within the B_U and B_L boundaries in Fig. 5 as shown in Fig. 6.

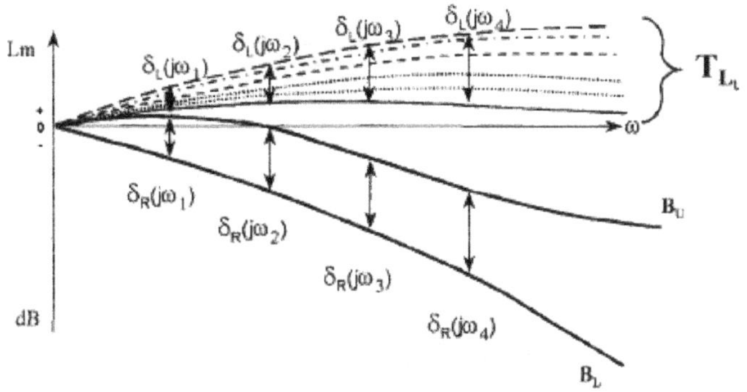

Fig. 5. Closed-loop responses: LTI plants with G(s)

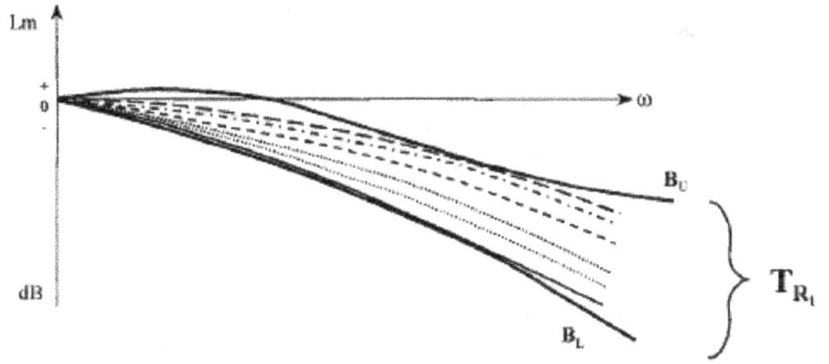

Fig. 6. Closed-loop responses: LTI plants with G(s) and F(s)

Therefore, the QFT robust design technique assures that the desired performance specifications are satisfied over the prescribed region of structured plant parametric uncertainty.

3. Insight to the QFT technique

3.1 Open-loop plant

Consider a certain position control system whose plant transfer function is given by

$$P_i(s) = \frac{K_a}{s(s+a)} = \frac{K'}{s(s+a)} \tag{1}$$

where $K' = K_a$ and $i = 1,2,...,J$. The log magnitude changes in a prescribed range due to the plant parameter uncertainty. The loop transmission $L(s)$ is defined as

$$L_i(s) = G(s)P_i(s) \tag{2}$$

3.2 Closed-loop formulation

The control ratio T_L of the unity-feedback system of Fig. 2 is

$$T_{Lt} = \frac{Y}{R_L} = \frac{L_t}{1+L_t} \tag{3}$$

The overall system control ratio T_R

is given by:

$$T_{Rt}(s) = \frac{F(s)L_t(s)}{1+L_t(s)} \tag{4}$$

3.3 Results of applying the QFT design technique

The proper application of the robust QFT design technique requires the utilization of the prescribed performance specifications from the onset of the design process, and the selection of a nominal plant P_o from the J LTI plants. Once the proper loop shaping of $L_o(s) = G(s)P_o(s)$ is accomplished, a synthesized $G(s)$ is achieved that satisfies the desired performance specifications. The last step of this design process is the synthesis of the prefilter that ensures that the Bode plots of T_{Ri} all lie between the upper and lower bounds B_U and B_L .

3.4 Benefits of QFT

The benefits of the QFT technique may be summarized as follows:

- It results in a robust design which is insensitive to structured plant parameter variation.
- There can be one robust design for the full, operating envelope.
- Design limitations are apparent up front and during the design process.
- The achievable performance specifications can be determined in the early design stage.
- If necessary, one can redesign for changes in the specifications quickly with the aid of the QFT CAD package.
- The structure of the compensator (controller) is determined up front.
- There is less development time for a full envelope design.

4. QFT design for the MISO analog control system

4.1 Introduction

The MIMO synthesis problem is converted into a number of single-loop feedback problems in which parameter uncertainty, cross-coupling effects, and system performance tolerances are derived from the original MIMO problem. The solutions to these single-loop problems represent a solution to the MIMO plant. It is not necessary to consider the complete system characteristic equation. The design is tuned to the extent of the uncertainty and the performance tolerances.

Here, we will present an in-depth understanding and appreciation of the power of the QFT technique through apply QFT to a robust single-loop MISO system, which has two inputs, a tracking and an external disturbance input, respectively, and a single output control system.

4.2 The QFT method (single-loop MISO system)

Basic structure of a feedback control system is given in Fig.7 , in which φ represents the set of transfer functions which describe the region of plant parameter uncertainty, G is the cascade compensator, and F is an input prefilter transfer function. The output $y(t)$ is required to track the command input $r(t)$ and to reject the external disturbances $d_1(t)$ and $d_2(t)$. The compensator G in Fig. 7 is to be designed so that the variation of $y(t)$ to the uncertainty in the plant P is within allowable tolerances and the effects of the disturbances $d_1(t)$ and $d_2(t)$ on $y(t)$ are acceptably small. Also, the prefilter properties of $F(s)$ must be designed to the desired tracking by the output $y(t)$ of the input $r(t)$. Since the control system in Fig. 7 has two measurable quantities, $r(t)$ and $y(t)$, it is referred to as a two degree-of-freedom (DOF) feedback structure. If the two disturbance inputs are measurable, then it represents a four DOF structure. The actual design is closely related to the extent of the uncertainty and to the narrowness of the performance tolerances. The uncertainty of the plant transfer function is denoted by the set

$$\varphi = \{P_t\} \quad where\ t = 1,2,...,J \tag{5}$$

and is illustrated as follows.

Given that the plant transfer function is

$$P(s) = \frac{K}{s(s+a)} \tag{6}$$

where the value of K is in the range [1, 10] and a is in the range [-2, 2]. The design objective is to guarantee that $T_R(s) = Y(s)/R(s)$ and $T_D(s) = Y(s)/D(s)$ are members of the sets of acceptable $\Im_R$ and $\Im_D$ for changes of K and a . In a feedback control system, the principal challenge in the control system design is to relate the system performance specifications to the requirements on the loop transmission function $L(s) = G(s)P(s)$ in order to achieve the desired benefits of feedback, i.e., the desired reduction in sensitivity to plant uncertainty and desired disturbance attenuation. The advantage of the frequency domain is that $L(s) = G(s)P(s)$ is simply the multiplication of complex numbers. In the frequency domain it is possible to evaluate $L(j\omega)$ at every ω_i separately, and thus, at each ω_i , the optimal bounds on $L(j\omega)$ can be determined.

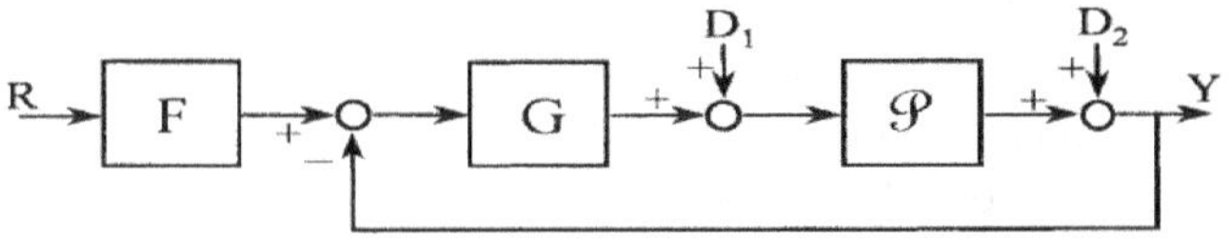

Fig. 7. A feedback structure

4.3 QFT design procedure

The objective is to design the prefilter $F(s)$ and the compensator $G(s)$ of Fig.7 so that the specified robust design is achieved for the given region of plant parameter uncertainty. The design procedure to accomplish this objective is as follows:

Step 1. Synthesize the desired tracking model.
Step 2. Synthesize the desired disturbance model.
Step 3. Specify the J LTI plant models that define the boundary of the region of plant parameter uncertainty.
Step 4. Obtain plant templates at specified frequencies that pictorially describe the region of plant parameter uncertainty on the Nichols chart.
Step 5. Select the nominal plant transfer function $P_o(s)$.
Step 6. Determine the stability contour (U -contour) on the Nichols chart.
Step 7-9. Determine the disturbance, tracking, and optimal bounds on the Nichols chart.
Step 10. Synthesize the nominal loop transmission function $L_o(s) = G(s)P_o(s)$ that satisfies all the bounds and the stability contour.
Step 11. Based upon Steps 1 through 10, synthesize the prefilter $F(s)$.
Step 12. Simulate the system in order to obtain the time response data for each of the J plants.

The following sections will illustrate the design procedure step by step.

4.4 Minimum-phase system performance specifications

In order to apply the QFT technique, it is necessary to synthesize the desired model control ratio based upon the system's desired performance specifications in the time domain. For the minimum-phase LTI MISO system of Fig. 7, the control ratios for tracking and for disturbance rejection are, respectively,

$$T_R(s) = \frac{F(s)G(s)P(s)}{1+G(s)P(s)} = \frac{F(s)L(s)}{1+L(s)} = F(s)T(s) \quad \text{with } d_1(t) = d_2(t) = 0 \tag{7}$$

$$T_{D1} = \frac{P(s)}{1+G(s)P(s)} = \frac{P}{1+L} \quad \text{with } r(t) = d_2(t) = 0 \tag{8}$$

$$T_{D2} = \frac{1}{1+G(s)P(s)} = \frac{1}{1+L} \quad \text{with } r(t) = d_1(t) = 0 \tag{9}$$

4.4.1 Tracking models

The QFT technique requires that the desired tracking control ratios be modeled in the frequency domain to satisfy the required gain K_m and the desired time domain performance specifications for a step input. Thus, the system's tracking performance specifications for a simple second-order system are based upon satisfying some or all of the step forcing function figures of merit (FOM) for under-damped $(M_p, t_p, t_s, t_r, K_m)$ and over-damped (t_s, t_r, K_m) responses, respectively. These are graphically depicted in Fig. 8. The time responses $y(t)_U$ and $y(t)_L$ in this figure represent the upper and lower bounds, respectively, of the tracking performance specifications; that is, an acceptable response $y(t)$ must lie between these bounds. The Bode plots of the upper bound B_U and lower bound B_L for $Lm\, T_R(j\omega)$ vs. ω are shown in Fig. 9.

It is desirable to synthesize the control ratios corresponding to the upper and lower bounds T_{RU} and T_{RL}, respectively, so that $\delta_R(j\omega_i)$ increases as ω_i increases above the 0-dB crossing frequency ω_{cf} (see Fig. 9b) of T_{RU}. This characteristic of $\delta_R(j\omega_i)$ simplifies the process of synthesizing the loop transmission $L_o(s) = G(s)P_o(s)$ as discussed in Sec. 4.13 of this chapter. To synthesize $L_o(s)$, it is necessary to determine the tracking bounds $B_R(j\omega_i)$ (see Sec. 4.9) which are obtained based upon $\delta_R(j\omega_i)$. This characteristic of $\delta_R(j\omega_i)$ ensures that the tracking bounds $B_R(j\omega_i)$ decrease in magnitude as ω_i increases.

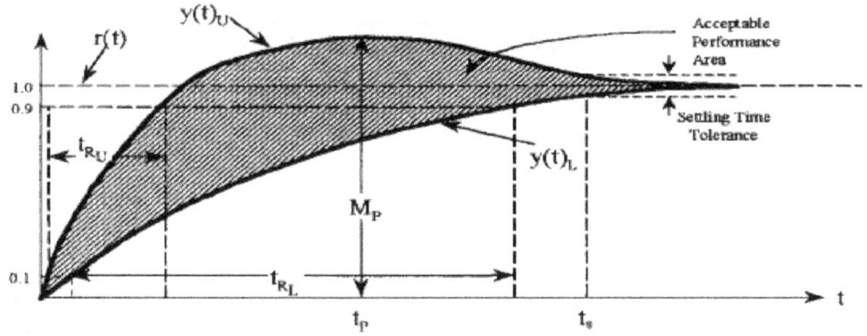

Fig. 8. System time domain tracking performance specifications

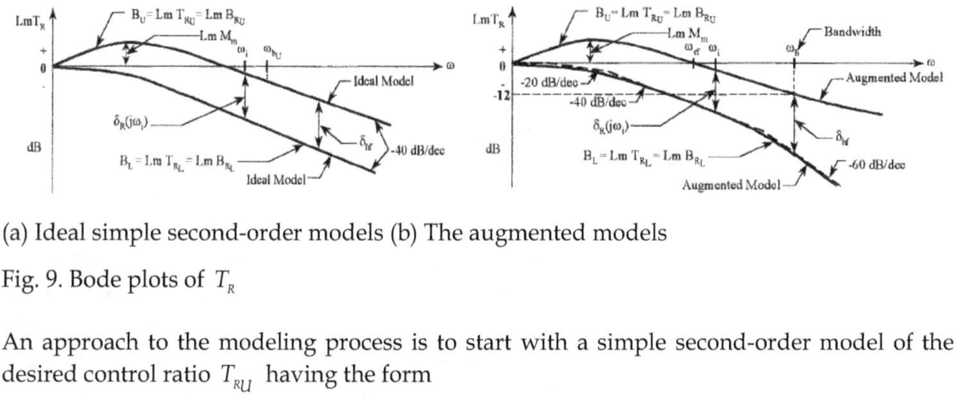

(a) Ideal simple second-order models (b) The augmented models

Fig. 9. Bode plots of T_R

An approach to the modeling process is to start with a simple second-order model of the desired control ratio T_{RU} having the form

$$T_{RU}(s) = \frac{\omega_n^2}{s^2 + 2\varsigma\omega_n s + \omega_n^2} = \frac{\omega_n^2}{(s - p_1)(s - p_2)} \tag{10}$$

where $\omega_n^2 = p_1 p_2$ and $t_s \approx T_s = 4/\varsigma\omega_n = 4/|\sigma_D|$ (the desired settling time). The control ratio $T_{RU}(s)$ of Eq. (10) can be represented by an equivalent unity-feedback system so that

$$T_{RU}(s) = \frac{Y(s)}{R(s)} = \frac{G_{eq}(s)}{1 + G_{eq}(s)} \tag{11}$$

where

$$G_{eq}(s) = \frac{\omega_n^2}{s(s+2\varsigma\omega_n)} \tag{12}$$

The gain constant of this equivalent Type1 transfer function $G_{eq}(s)$ is $K_1 = \lim_{s\to 0}[sG_{eq}(s)] = \omega_n / 2\varsigma$.

The simplest over-damped model for $T_{RL}(s)$ is of the form

$$T_{RL}(s) = \frac{Y(s)}{R(s)} = \frac{K}{(s-\sigma_1)(s-\sigma_2)} = \frac{G_{eq}(s)}{1+G_{eq}(s)} \tag{13}$$

where

$$G_{eq}(s) = \frac{\sigma_1\sigma_2}{s[s-(\sigma_1+\sigma_2)]}$$

and $K_1 = -\sigma_1\sigma_2 / (\sigma_1+\sigma_2)$. Selection of the parameters σ_1 and σ_2 is used to meet the specifications for t_s and K_1.

Once the ideal models $T_{RU}(j\omega)$ and $T_{RL}(j\omega)$ are determined, the time and frequency response plots of Figs. 8 and 9a, respectively, can then be drawn. The high-frequency range in Fig. 9a is defined as $\omega \geq \omega_b$, where ω_b is the model BW frequency of B_U. In order to achieve the desired characteristic of an increasing magnitude of δ_R of B_U for $\omega_i > \omega_{cf}$, an increasing spread between B_u and B_L is required in the high-frequency range (see Fig. 9b), that is,

$$\delta_{hf} = B_U - B_L \tag{14}$$

must increase with increasing frequency. This desired increase in δ_R is achieved by changing B_u and B_L by augmenting T_{RU} with a zero [see Eq. (15)] as close to the origin as possible without significantly affecting the time response. This additional zero raises the curve B_u for the frequency range above ω_{cf}. The spread can be further increased by augmenting T_{RL} with a negative real pole [see Eq. (16)] which is as close to the origin as possible but far enough away not to significantly affect the time response. Note that the straight-line Bode plot is shown only for T_{RL}. This additional pole lowers B_L for this frequency range.

$$T_{RU}(s) = \frac{(\omega_n^2/a)(s+a)}{s^2+2\varsigma\omega_n s+\omega_n^2} = \frac{(\omega_n^2/a)(s-z_1)}{(s-\sigma_1)(s-\sigma_2)} \tag{15}$$

$$T_{RL}(s) = \frac{K}{(s+a_1)(s+a_2)(s+a_3)} = \frac{K}{(s-\sigma_1)(s-\sigma_2)(s-\sigma_3)} \tag{16}$$

Thus, the magnitude of $\delta_R(j\omega_i)$ increases as ω_i, increases above ω_{cf}.

In order to minimize the iteration process in achieving acceptable models for $T_{RU}(s)$ and $T_{RL}(s)$ which have an increasing $\delta_R(j\omega)$, the following procedure may expedite the design process: (a) first synthesize the second-order model of Eq. (15) containing the zero at $|z_1| = a \geq \omega_n$ that meets the desired FOM; and (b) then, as a first trial, select all three real poles of Eq. (16) to have the value of $|\sigma_3| = a_3 = \omega_n > a_2 = a_1 > |\sigma_D|$. For succeeding trials, if necessary, one or more of these poles are moved right and/or left until the desired specifications are satisfied. As illustrated by the slopes of the straight-line Bode plots in Fig. 9b, selecting the value of all three poles in the range specified above insures an increasing δ_R. Other possibilities are as follows: (c) the specified values of t_p and t_s for T_{RL} may be such that a pair of complex poles and a real pole need to be chosen for the model response. For this situation, the real pole must be more dominant than the complex poles, (d) depending on the performance specifications, $T_{RU}(s)$ may require two real poles and a zero "close" to the origin, i.e., select $|z_1|$ very much less than $|p_1|$ and $|p_2|$ in order to effectively have an under-damped response.

At high frequencies δ_{hf} (see Fig. 9b) must be larger than the actual variation in the plant, δ_P. For the case where $y(t)$, corresponding to T_{RU}, is to have an allowable "large" overshoot followed by a small tolerable undershoot, a dominant complex pole pair is not suitable for T_{RU}. An acceptable overshoot with no undershoot for T_{RU} can be achieved by T_{RU} having two real dominant poles $p_1 > p_2$, a dominant real zero ($z_1 > p_1$) "close"' to p_1, and a far off pole $p_3 \ll p_2$. The closeness of the zero dictates the value of M_p. Thus, a designer selects a pole-zero combination to yield the form of the desired time-domain response.

4.4.2 Disturbance rejection models

The simplest disturbance control ratio model specification is $|T_D(j\omega)| = |Y(j\omega / D(j\omega))| < a_p$, a constant, [the desired maximum magnitude of the output based upon a unit-step disturbance input]; i.e., for $d_1(t): |y(t_p)| \leq a_p$, and for: $d_2(t)|y(t)| \leq a_p$ for $t \geq t_x$. Thus, the frequency domain disturbance specification is $Lm\, T_D(j\omega) < Lm\, a_p$ over the desired specified BW (see Fig. 10).

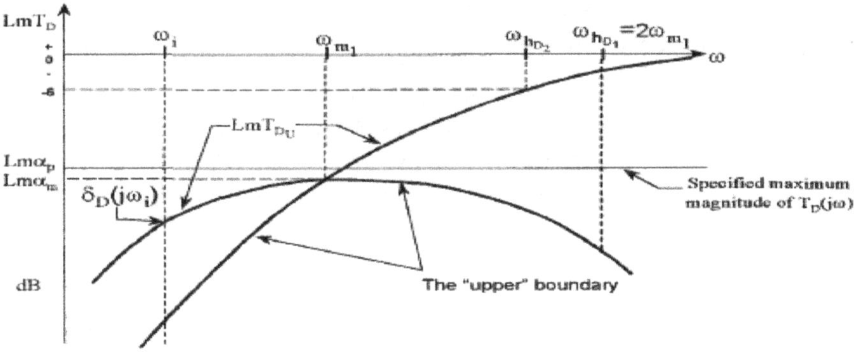

Fig. 10. Bode plots of disturbance models for $T_D(j\omega)$

4.5 *J* LTI plant models

The simple plant of Eq. (17)

$$P_i(s) = \frac{Ka}{s(s+a)} \tag{17}$$

where $K \in \{1,10\}$ and $a \in \{1,10\}$, is used to illustrate the MISO QFT design procedure. The region of plant parameter uncertainty may be described by *J* LTI plants, where $i = 1,2,\ldots,J$ which lie on its boundary.

4.6 Plant templates of $P_i(s), \Im P(j\omega_i)$

With $L = GP$, Eq. (7) yields

$$Lm\ T_R = Lm\ F + Lm\left[\frac{L}{1+L}\right] = Lm\ F + Lm\ T \tag{18}$$

The change in T_R due to the uncertainty in P, since F is LTI, is

$$\Delta(Lm\ T_R) = Lm\ T_R - Lm\ F = Lm\left[\frac{L}{1+L}\right] \tag{19}$$

The proper design of $L = L_o$ and F, must restrict this change in T_R so that the actual value of $Lm\ T_R$ always lies between B_U and B_L of Fig. 9b. The first step in synthesizing an L_o is to make NC templates which characterize the variation of the plant uncertainty for various values of ω_i, over a frequency range $\omega_x \le \omega_i \le \omega_{hR}$, where $\omega_x \le \omega_{cf}$. For the plant of Eq. (17) the values K = a = 1 represent the lowest point of each of the templates $\Im P(j\omega_i)$ and may be selected as the nominal plant P_o for all frequencies.

4.7 Nominal plant

While any plant case can be chosen, it is a common practice to select, whenever possible, a plant whose NC point is always at the lower left corner of the template for all frequencies for which the templates are obtained.

4.8 *U*-contour (stability bound)

The specifications on system performance in the time domain (see Fig. 8) and in the frequency domain (see Fig. 9) identify a minimum damping ratio ς for the dominant roots of the closed-loop system which corresponds to a bound on the value of $M_p \approx M_m$. On the NC this bound on $M_p \approx M_L$ (see Fig. 11) establishes a region which must not be penetrated by the templates and the loop transmission functions $L_i(j\omega)$ for all ω. The boundary of this region is referred to as the stability bound, the U-contour, because this becomes the dominating constraint on $L(j\omega)$. Therefore, the top portion, indicated by the coordinates efa, of the M_L contour becomes part of the U-contour. The formation of the U -contour is discussed in this section. For the two cases of disturbance rejection depicted in Fig. 7 the control ratios are, respectively, as given in Eqs. (8) and (9).

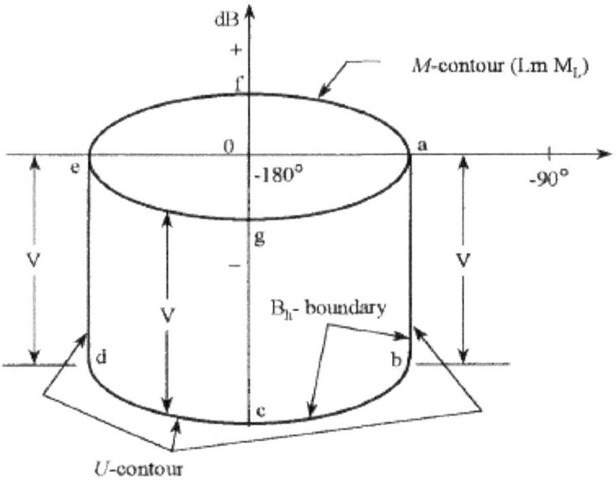

Fig. 11. U-contour construction (stability contour)

Thus, it is necessary to synthesize an $L_o(s)$ so that the disturbances are properly attenuated. For the present, only one aspect of this disturbance-response problem is considered, namely a constraint is placed on the damping ratio ς of the dominant complex-pole pair of T_D nearest the ω-axis. This damping ratio is related to the peak value of

$$|T(j\omega)| = \left| \frac{L(j\omega)}{1 + L(j\omega)} \right| \tag{20}$$

Therefore, it is reasonable to add the requirement

$$|T| = \left| \frac{L}{1 + L} \right| \le M_L \tag{21}$$

where M_L is a constant for all ω and over the whole range of φ parameter values. This results in a constraint on ς of the dominant complex-pole pair of T_D. This constraint can therefore be transformed into a constraint on the maximum value $T_{\max}$ of Eq. (20). This results in limiting the peak of the disturbance response. A value of M_L can be selected to correspond to the maximum value of T_R. Therefore, the top portion, efa as shown in Fig.11, of the M-contour on the NC, which corresponds to the value of the selected value of M_L, becomes part of the U-contour.

For a large class of problems, as $\omega \to \infty$, the limiting value of the plant transfer function approaches

$$\lim_{\omega \to \infty} [P(j\omega)] = \frac{K'}{\omega^{\lambda}}$$

where λ represents the excess of poles over zeros of $P(s)$.

4.9 Tracking bounds $B_R(j\omega_i)$

Consider the plot of $Lm\,P(j\omega)$ vs. $\angle P(j\omega)$ for a plant shown in Fig. 12 (the solid curve). With $G(s) = A = 1$ and $F(s) = 1$ in Fig. 7, $L = P$. The plot of $Lm\,L(j\omega)$ vs. $\angle L(j\omega)$ is tangent to the M = 1dB curve with a resonant frequency $\omega_m = 1.1$. If $Lm\,M_m = 2$ dB is specified for $Lm\,T_R$, the gain A is increased, raising $Lm\,L(j\omega)$, until it is tangent to the 2-dB M-curve. For this example the curve is raised by $Lm\,A = 4.5dB(G = A = 1.679)$ and the resonant frequency is $\omega_m = 2.09$.

Now consider that the plant uncertainty involves only the variation in gain A between the values of 1 and 1.679. It is desired to find a cascade compensator $G(s)$, in Fig. 7, such that the specification $1\,dB < Lm\,M_m < 2\,dB$ is always maintained for this plant gain variation while the resonant frequency ω_m remains constant. This requires that the loop transmission $L(j\omega) = G(j\omega)P(j\omega)$ be synthesized so that it is tangent to an M-contour in the range of $1\,dB < Lm\,M < 2\,dB$ for the entire range of $1 < A < 1.679$ and the resultant resonant frequency satisfies the requirement $\omega_m = 2.09 + \Delta\omega_m$.

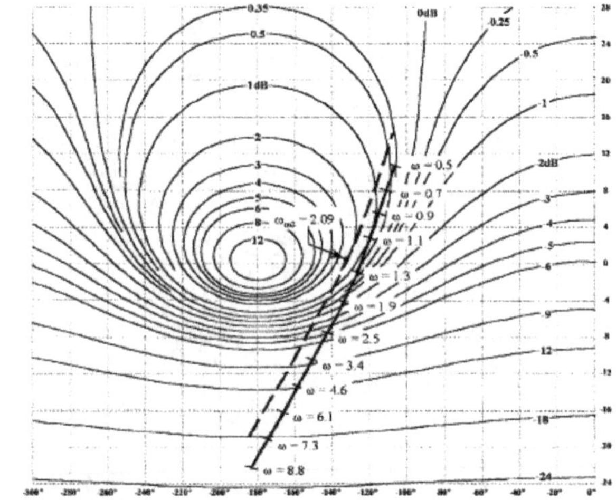

Fig. 12. Log magnitude-angle diagram

It is assumed for Eq. (19) that the compensators F and G are fixed (LTI), that is, they have negligible uncertainty. Thus, only the uncertainty in P contributes to the change in T_R given by Eq. (19). The solution requires that the actual $\Delta LmT_R(j\omega_i) \le \delta_R(j\omega_i)$ dB in Fig. 9b. Thus, it is necessary to determine the resulting constraint, or bound $B_R(j\omega_i)$, on $L(j\omega_i)$. The procedure is to select a nominal plant $P_o(s)$ and to derive the bounds on the resulting nominal loop transfer function $L_o(s) = G(s)P_o(s)$.

As an illustration, consider the plot of $Lm\,P(j2) vs. \angle P(j2)$ for the plant of Eq. (17). As shown in Fig. 13, the plant's region of uncertainty $\Im P(j2)$ is given by the contour ***ABCD***, i.e., $Lm\,P(j2)$ lies on or within the boundary of this contour. The nominal plant transfer function, with $K_o = 1$ and $a_o = 1$, is

$$P_o(s) = \frac{1}{s(s+1)} \tag{22}$$

and is represented in Fig. 13 by point A for $\omega = 2$ [-13.0 dB, -153.4°}. Note, once a nominal plant is chosen, it must be used for determining all the bounds $B_R(j\omega_i)$.

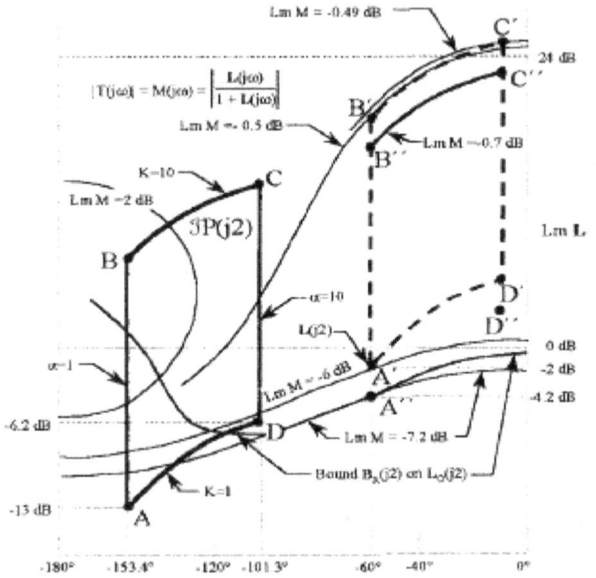

Fig. 13. Derivation of bounds $B_R(j\omega_i)$ on $L_o(j\omega)$ for $\omega = 2$

4.10 Disturbance bounds $B_D(j\omega_i)$: CASE 1

Two disturbance inputs are shown in Fig. 7. It is assumed that only one disturbance input exists at a time. Both cases are analyzed.

CASE 1 [$d_2(t) = D_0 u_{-1}(t), d_1(t) = 0$]

CONTROL RATIO. From Fig. 7 , the disturbance control ratio for input $d_2(t)$ is

$$T_D(s) = \frac{1}{1+L} \tag{23}$$

Substituting $L = 1/\ell$ into Eq. (23) yields

$$T_D(s) = \frac{\ell}{1+\ell} \tag{24}$$

this equation has the mathematical format required to use the NC. Over the specified BW it is desired that $|T_D(j\omega)| \ll 1$, which results in the requirement, from Eq.(24), that $|L(j\omega)| \gg 1$, i.e.,

$$T_D(j\omega) \approx \frac{1}{|L(j\omega)|} = |\ell(j\omega)|$$

DISTURBANCE RESPONSE CHARACTERISTIC. A time-domain tracking response characteristic based upon $r(t) = u_{-1}(t)$ often specifies a maximum allowable peak overshoot M_p. In the frequency domain this specification may be approximated by

$$|M_R(j\omega)| = |T_R(j\omega)| = \left|\frac{Y(j\omega)}{R(j\omega)}\right| \le M_m \approx M_P \tag{25}$$

The corresponding time- and frequency-domain response characteristics, based upon the step disturbance forcing function $d_2(t) = u_{-1}(t)$, are, respectively,

$$|M_D(t)| = \left|\frac{Y(t)}{d(t)}\right| \le \alpha_P \quad for\ t \ge t_x \tag{26}$$

and

$$|M_D(j\omega)| = |T_D(j\omega)| = \left|\frac{Y(j\omega)}{D(j\omega)}\right| \le \alpha_m \approx \alpha_P \tag{27}$$

4.11 Disturbance bounds $B_D(j\omega_i)$: CASE 2

CASE 2 [$d_1(t) = D_0 u_{-1}(t), d_2(t) = 0$]

CONTROL RATIO. From Fig. 7, the disturbance control ratio for the input $d_1(t)$ is

$$T_D(j\omega) = \frac{P(j\omega)}{1 + G(j\omega)P(j\omega)} \tag{28}$$

Assuming point A of the template represents the nominal plant P_o. Eq. (28) is multiplied by P_o / P_o and rearranged as follows:

$$T_D = \frac{P_o}{P_o}\left[\frac{1}{\frac{1}{P} + G}\right] = \frac{P_o}{\frac{P_o}{P} + GP_o} = \frac{P_o}{\frac{P_o}{P} + L_o} = \frac{P_o}{W} \tag{29}$$

where

$$W = (P_o / P) + L_o \tag{30}$$

Thus Eq.(29) with $Lm\, T_D = \delta_D$ yields

$$Lm\, W = Lm\, P_o - \delta_D \tag{31}$$

DISTURBANCE RESPONSE CHARACTERISTICS. Based on Eq. (25), the time and frequency-domain response characteristics, for a unit-step disturbance forcing function, are given, respectively, by

$$|M_D(t)| = \left|\frac{y(t_p)}{d(t)}\right| = |y(t_p)| \le \alpha_p \tag{32}$$

and

$$|M_D(j\omega)| = |T_D(j\omega)| = \left|\frac{Y(j\omega)}{D(j\omega)}\right| \le \alpha_m \equiv \alpha_p \tag{33}$$

where t_p is the peak time.

4.12 The composite boundary $B_o(j\omega_i)$

The composite bound $B_o(j\omega_i)$ that is used to synthesize the desired loop transmission transfer function $L_o(s)$ is obtained in the manner shown in Fig. 14. The composite bound $B_o(j\omega_i)$, for each value of ω_i, is composed of those portions of each respective bound $B_R(j\omega_i)$ and $B_D(j\omega_i)$ that are the most restrictive. For the case shown in Fig. 14a the bound $B_o(j\omega_i)$ is composed of those portions of each respective bound $B_R(j\omega_i)$ and $B_D(j\omega_i)$ that have the largest values. For the situation of Fig. 14b, the outermost of the two boundaries $B_R(j\omega_i)$ and $B_D(j\omega_i)$ becomes the perimeter of $B_o(j\omega_i)$. The situations of Fig. 14 occur when the two bounds have one or more intersections. If there are no intersections, then the bound with the largest value or with the outermost boundary dominates. The synthesized $L_o(j\omega_i)$, for the situation of Fig. 14a, must be on or just above the bound $B_o(j\omega_i)$. For the situation of Fig. 14b the synthesized $L_o(j\omega_i)$ must not lie in the interior of the $B_o(j\omega_i)$ contour.

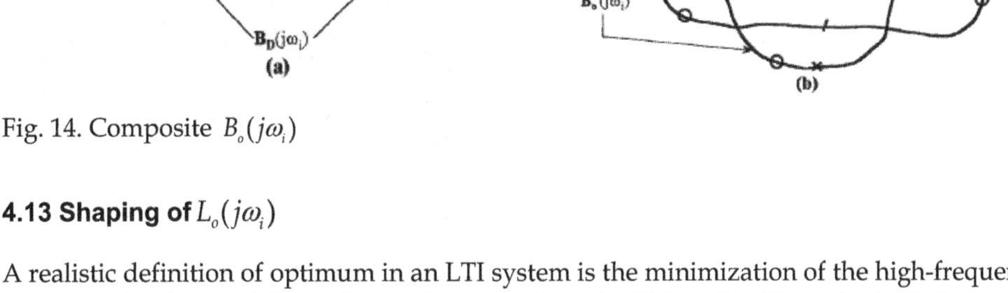

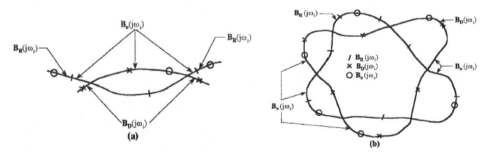

Fig. 14. Composite $B_o(j\omega_i)$

4.13 Shaping of $L_o(j\omega_i)$

A realistic definition of optimum in an LTI system is the minimization of the high-frequency loop gain K while satisfying the performance bounds. This gain affects the high-frequency response since $\lim_{\omega\to\infty}[L(j\omega)] = K(j\omega)^{-\lambda}$ where λ is the excess of poles over zeros assigned to $L(j\omega)$. Thus, only the gain K has a significant effect on the high-frequency response, and the effect of the other parameter uncertainty is negligible. Also, the importance of minimizing the high-frequency loop gain is to minimize the effect of sensor noise whose spectrum, in general, lies in the high-frequency range.

For the plant of Eq. (17), the shaping of $L_o(j\omega)$ is shown by the dashed curve in Fig. 15. A point such as $Lm\,L_o(j2)$ must be on or above the curve labeled $B_o(j2)$. Further, in order to satisfy the specifications, $L_o(j\omega)$ cannot violate the U-contour. In this example a reasonable $L_o(j\omega)$ closely follows the U-contour up to $\omega = 40$ rad/sec and stays below it above $\omega = 40$ as shown in Fig 15. Additional specifications are $\lambda = 4$, i.e., there are 4 poles in excess of zeros, and that it also must be Type 1 (one pole at the origin).A representative procedure for choosing a rational function $L_o(s)$ which satisfies the above specifications is now described. It involves building up the function

$$L_o(j\omega) = L_{ok}(j\omega) = P_o(j\omega)\prod_{k=0}^{w}[K_k G_k(j\omega)] \tag{34}$$

where for k = 0, $G_o = 1\angle 0^\circ$, and $K = \prod_{k=0}^{w} K_k$

In order to minimize the order of the compensator, a good starting point for "building up" the loop transmission function is to initially assume that $L_{o0}(j\omega) = P_o(j\omega)$ as indicated in Eq. (34). $L_o(j\omega)$ is built up term-by-term in order to stay just outside the U-contour in the NC of Fig. 15. The first step is to find the $B_o(j\omega_i)$ which dominates $L_o(j\omega)$.

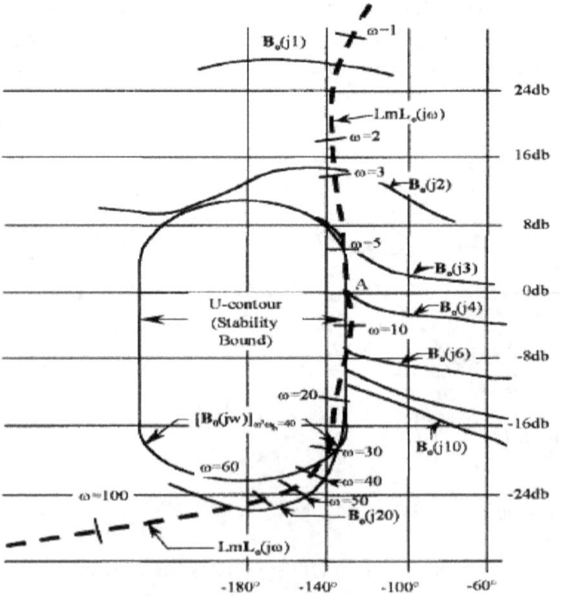

Fig. 15. Shaping of $L_o(j\omega)$ on the Nichols chart for the plant of Eq. (17)

4.14 Design of the prefilter $F(s)$

Design of a proper $L_o(s)$ guarantees only that the variation in $|T_R(j\omega)|$, i.e., ΔT_R, is less than or equal to that allowed. The purpose of the prefilter is to position $Lm\,T(j\omega)$ within the frequency domain specifications. For the example of this chapter the magnitude of the frequency response must be within the bounds B_U and B_L shown in Fig. 9b, which are redrawn in Fig. 16. A method for determining the bounds on $F(s)$ is as follows: Place the nominal point A of the ω_i plant template on the $L_o(j\omega_i)$ point of the $L_o(j\omega)$ curve on the NC (see Fig. 17). Traversing the template, determine the maximum $Lm\,T_{max}$ and minimum $Lm\,T_{min}$, values of

$$Lm\,T(j\omega_i) = \frac{L(j\omega_i)}{1 + L(j\omega_i)} \tag{35}$$

obtained from the M-contours. These values are plotted as shown in Fig. 16. The tracking control ratio is $T_R = FL/[1+L]$ and

$$Lm\,T_R(j\omega_i) = Lm\,F(j\omega_i) + Lm\,T(j\omega_i) \tag{36}$$

The variations in Eqs. (35) and (36) are both due to the variation in P; thus

$$\delta_L(j\omega_i) = Lm\,T_{max} - Lm\,T_{min} \le \delta_R = B_U - B_L \tag{37}$$

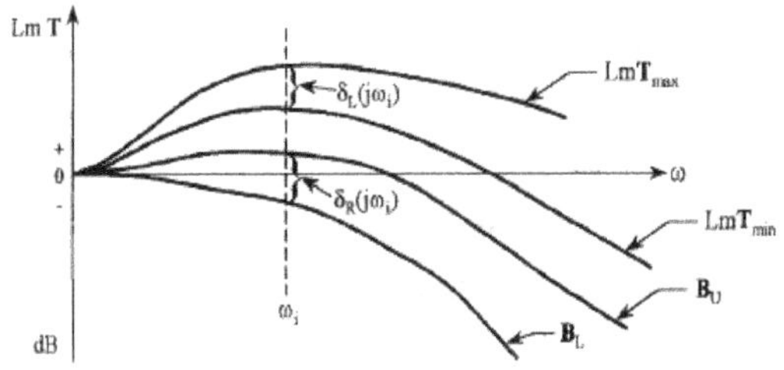

Fig. 16. Requirements on $F(s)$

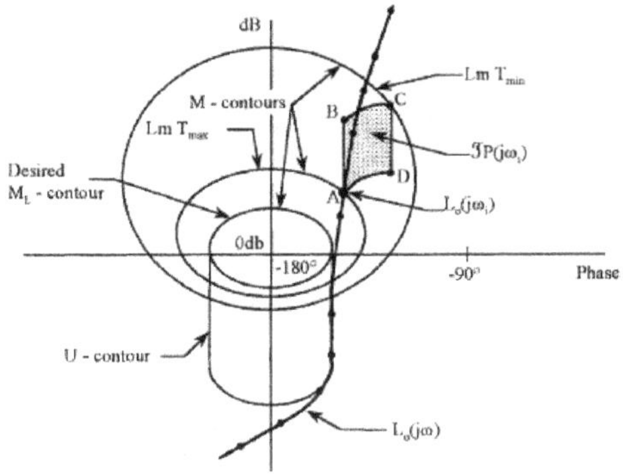

Fig. 17. Prefilter determination

If values of $L_o(j\omega)$, for each value ω_i, lie exactly on the tracking bounds $B_R(j\omega_i)$, then $\delta_L = \delta_R$. Therefore, based upon Eq. (36), it is necessary to determine the range in dB by which $Lm\,T(j\omega_i)$ must be raised or lowered to fit within the bounds of the specifications by use of the prefilter $F(j\omega_i)$. The process is repeated for each frequency corresponding to the templates used in the design of $L_o(j\omega)$. Therefore, in Fig. 18 the difference between the $Lm\,T_{RU} - Lm\,T_{max}$ and the $Lm\,T_{RL} - Lm\,T_{min}$ curves yields the requirement for $Lm\,F(j\omega)$, i.e., from Eq. (36).

$$Lm\,F(j\omega) = Lm\,T_R(j\omega) - Lm\,T(j\omega) \tag{38}$$

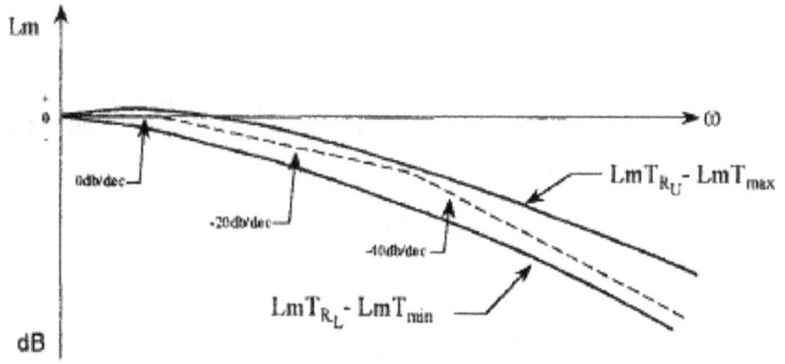

Fig. 18. Frequency bounds on the prefilter $F(s)$

The procedure for designing $F(s)$ is summarized as follows:

1. Use templates in conjunction with the $L_o(j\omega)$ plot on the NC to determine T_{max} and T_{min} for each ω_i. This is done by placing $\Im P(j\omega_i)$ with its nominal point on the point $Lm\,L_o(j\omega)$. Then use the M-contours to determine $T_{max}(j\omega_i)$ and $T_{min}(j\omega_i)$ (see Fig. 17).
2. Obtain the values of $Lm\,T_{RU}$ and $Lm\,T_{RL}$ for various values of a, from Fig. 9b.
3. From the values obtained in steps 1 and 2, plot $\left[Lm\,T_{RU} - Lm\,T_{max}\right]$ and $\left[Lm\,T_{RL} - Lm\,T_{min}\right]$ vs. ω as shown in Fig. 18.
4. Use straight-line approximations to synthesize an $F(s)$ so that $Lm\,F(j\omega_i)$ lies within the plots of step 3. For step forcing functions the resulting $F(s)$ must satisfy

$$\lim_{s\to 0}[F(s)] = 1 \tag{39}$$

4.15 Basic design procedure for a MISO system

The basic concepts of the QFT technique are explained by means of a design example. The system configuration shown in Fig. 7 contains three inputs. The first objectives are to track a step input $r(t) = u_{-1}(t)$ with no steady-state error and to satisfy the performance specifications of Fig. 8. An additional objective is to attenuate the system response caused by external step disturbance inputs $d_1(t)$ and $d_2(t)$. An outline of the basic design procedure for the QFT technique, as applied to a minimum-phase plant, is as follows:

1. Synthesize the tracking model control ratio $T_R(s)$ in the way described in Sec. 4.4, based upon the desired tracking specifications (see Figs. 8 and 9b).
2. Synthesize the disturbance-rejection model control ratios $T_D(s)$ in the manner described in Sec. 4.10 based upon the disturbance-rejection specifications.
3. Obtain templates of $P(j\omega_i)$ that pictorially describe the plant uncertainty on the Nichols chart for the desired pass-band frequency range.
4. Select a nominal plant from the set of Eq. (5) and denote it as $P_o(s)$.
5. Determine the U-contour based upon the specified values of $\delta_R(j\omega_i)$ for tracking, M_L for disturbance rejection, and V for the universal high frequency boundary (UHFB) in conjunction with steps 6 through 8.
6. Use the data of steps 2 and 3 and the values of $\delta_D(j\omega_i)$ (see Fig. 10) to determine the disturbance bound $B_D(j\omega_i)$ on the loop transmission $L_D(j\omega_i) = G(j\omega_i)P_o(j\omega_i)$. For minimum-phase systems this requires that the synthesized loop transmission $Lm\, L_D(j\omega_i)$ must be on or above the curve for $Lm\, B_D(j\omega_i)$ on the Nichols diagram (see Fig. 15 assuming $B_D = B_o$).
7. Determine the tracking bound $B_R(j\omega_i)$ on the nominal transmission $L_o(j\omega_i) = G(j\omega_i)P_o(j\omega_i)$, using the tracking model (step 1), the templates $P(j\omega_i)$ (step 3), the values of $\delta_R(j\omega_i)$ (see Fig. 9b), and M_L [see Eq.(21)]. For minimum-phase systems this requires that the synthesized loop transmission satisfy the requirement that $Lm\, L_o(j\omega_i)$ is on or above the curve for $Lm\, B_R(j\omega_i)$ on the Nichols diagram.
8. Plot curves of $Lm\, B_R(j\omega_i)$ versus $\phi_R = \angle B_R(j\omega_i)$ and $Lm\, B_D(j\omega_i)$ versus $\phi_d = \angle B_D(j\omega_i)$ on the same NC. For a given value of ω_i at various values of the angle ϕ, select the value of $Lm\, B_D(j\omega_i)$ or $Lm\, B_R(j\omega_i)$, whichever is the largest value (termed the "worst" or "most severe" boundary). Draw a curve through these points. The resulting plot defines the overall boundary $Lm\, B_o(j\omega_i) vs. \phi$. Repeat this procedure for sufficient values of ω_i.
9. Design $L_o(j\omega_i)$ to be as close as possible to the boundary value $B_o(j\omega_i)$ by selecting an appropriate compensator transfer function $G(j\omega)$. Synthesize an $L_o(j\omega) = G(j\omega)P_o(j\omega)$ using the $Lm\, B_o(j\omega_i)$ boundaries and U-contour so that $Lm\, L_o(j\omega_i)$ is on or above the curve for $Lm\, B_o(j\omega_i)$ on the Nichols diagram.
10. Based upon the information available from steps 1 and 9, synthesize an $F(s)$ those results in a $Lm\, T_R$ [Eq. (7)] vs. ω that lies between B_U and B_L of Fig. 9b.
11. Obtain the time-response data for $y(t)$: (a) with $d(t) = u_{-1}(t)$ and $r(t) = 0$ and (b) with $r(t) = u_{-1}(t)$ and $d(t) = 0$ for sufficient points around the parameter space describing the plant uncertainty.

5. Robust QFT flight control design for a certain UAV

5.1 Introduction

Unmanned Aerial Vehicles (hereafter referred as UAVs) play a very important role in modern war. Whereas flight stability of UAVs is easily affected by airflow, model perturbation and other uncertainty. To enhance flight stability and robustness of UAVs,

$H\infty$ control, QFT technique, linear quadratic Gaussian (LQG) have been applied to UAVs' flight control system at present. Comparatively, QFT can take uncertainty's scopes and performance requirements into account, analyze and design robust controller on Nichols chart quantitatively in order to make the open-loop frequency curve comply with boundary conditions and have robust stability and performance robustness.

QFT has been widely used in aerospace field and is mature for robust controller design of LTI/SISO system. But QFT design for MIMO system still faces many difficulties. In view of the characteristics of a certain small UAV which used in tracking and surveillance, a novel QFT controller design method for the UAV's lateral motion is introduced in this section.

5.2 QFT design for MIMO systems

5.2.1 Overview

The QFT design for MIMO systems is based upon the mathematical means which results in the representation of a MIMO control system by m^2 MISO equivalent control systems. The highly structured uncertain LTT MIMO plant has the following features:

1. The synthesis problem is converted into a number of single-loop problems, in which structured parameter uncertainty, external disturbance, and performance tolerances are derived from the original MIMO problem. The solutions to these single-loop problems are guaranteed to work for the MIMO plant. It is not necessary to consider the system characteristic equation.
2. The design is tuned to the extent of the uncertainty and the performance tolerances. The design for a MIMO system, as stated previously, involves the design of an equivalent set of MISO system feedback loops.

The design process for these individual loops is the same as the design of a MISO system described in previous sections.

Pure mathematical transformation method used in QFT design for MIMO systems tends to cause a larger super-margin design and is very complicated when system is of higher order. Comparatively, Basically Non-interacting (hereafter referred as BNIA) is commonly used in practical applications. Note that principle of BNIA, which will be negligible here, can be found in references of this chapter.

5.2.2 Non-interacting (BNIA) loops

A BNIA loop is one in which the output $y_K(s)$ due to the input $r_j(s)$ is ideally zero. Plant uncertainty and loop interaction (cross-coupling) makes the ideal response unachievable. Thus, the system performance specifications describe a range of acceptable responses for the commanded output and a maximum tolerable response for the uncommanded outputs. The uncommanded outputs are treated as cross-coupling effects.

For an LTI plant having no parameter uncertainty, it is possible to essentially achieve zero cross-coupling effects, i.e., the output $y_K \approx 0$. This desired result can be achieved by post multiplying P by a matrix W to yield:

$$P_n = PW = [p_{ijn}] \quad where\; p_{ijn} = 0 \; for\; i \neq j$$

resulting in a diagonal P_n matrix for P representing the nominal plant case in the set φ. With plant uncertainty the off-diagonal terms of P_n will not be zero but "very small" in comparison to P, for the nonnominal plant cases in φ. In some design problems it may be necessary or desired to determine a P_n upon which the QFT design is accomplished. Doing this minimizes the effort required to achieve the desired BW and minimizes the cross-coupling effects.

5.3 QFT design and simulation for a certain UAV's lateral motion

QFT approach for MIMO system will be applied to a certain UAV's lateral motion in this section.

5.3.1 Mathematical model of the UAV

State equation of the UAV is generally expressed as:

$$\begin{cases} \dot{x}(t) = Ax(t) + Bu(t) \\ y(t) = Cx(t) \end{cases} \tag{40}$$

where $X = [\beta \quad p \quad \gamma \quad \phi \quad \delta_a \quad \delta_r]^T ; u = [\delta_{rc} \quad \delta_{ac}]^T ; Y = [\beta \quad p \quad \gamma \quad \phi]^T$; β is sideslip angle, p is roll angle rate, γ is yaw angle rate, ϕ is roll angle, δ_a is aileron deflection angle, δ_{rc} is rudder deflection angle, δ_{ac} is rudder deflection angle command input, A, B, C are system matrix, input matrix and input-output matrix respectively. By way of wind tunnel test and mathematic method, matrices A, B and C in eqs.(40) for the small UAV can be derived.

5.3.2 System decomposition

The UAV's lateral state equation described in Eq.(40) has two inputs and four outputs. According to QFT approach for MIMO system, we decompose Eq.(40) into two MISO subsystems using BNIA, one is yaw loop (loop I) subsystem, the other is roll loop (loop II) subsystem. QFT control structures of both loops are given in Fig.19 and Fig.20.

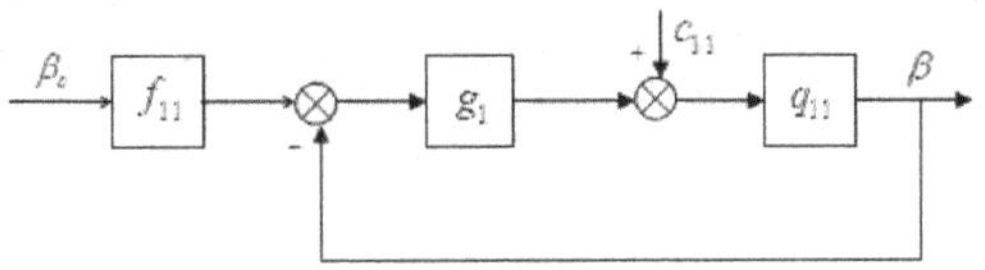

Fig. 19. QFT control structure of loop I

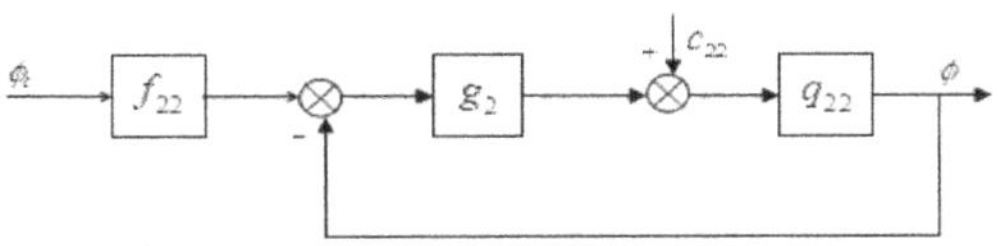

Fig. 20. QFT control structure of loop II

where β_c, ϕ_c are sideslip angle input and roll angle input respectively; g_1, g_2 are QFT controllers; f_{11}, f_{22} are QFT prefilters; c_{11}, c_{22} are disturbance inputs; q_{11}, q_{22} is controlled plants.

Decomposed state equation has relationship with that of the original system as follows:

$$A_c = A, B_c = B, C_c = \begin{bmatrix} 1 & 0 & 0 & 0 \\ 0 & 0 & 0 & 1 \end{bmatrix} C$$

Transfer function matrices P of decomposed plant can be easily derived as

$$P = C_c(sI - A_c)^{-1}B_c = \begin{bmatrix} p_{11} & p_{12} \\ p_{21} & p_{22} \end{bmatrix}$$

where p_{11} is the transfer function from δ_{rc} to β; p_{22} represents the transfer function from δ_{ac} to ϕ; p_{12} is the transfer function from δ_{rc} to ϕ, p_{21} represents the transfer function from δ_{ac} to β.

Next, we adopt 5 flight states to develop the QFT controllers of both loops.

5.3.3 QFT design for loop I

For loop I, we ensure $g_1(s)$ and $f_{11}(s)$ meet requirements of robust stability when β_c acts as command input and c_{11} as disturbance input. Besides, both subsystems should own ideal tracking performance and preferable noise restraint capability.

1. **Selection of Performance Indices.** Tracking performances indices of sideslip angle are overshoot $\sigma\% \le 2\%$, settling time $t_s \le 6\%$. Given the original model of upper tracking boundary is

$$T_{RU}(j\omega) = \frac{\omega_n^2}{s^2 + 2\varsigma\omega_n s + \omega_n^2} \tag{41}$$

According to $\sigma\%$ and t_s, damping ratio ς and natural oscillation frequency ω_n is adopted as 0.78 and 0.8978. Add a zero (z=-1) as close to the origin as possible without significantly affecting the time response(see Sec.4.4.1). This additional zero raises tracking boundary curve above ω_{cf}, the final transfer function of tracking curve's upper boundary is

$$T_{RU}(j\omega) = \frac{0.806(s+1)}{s^2 + 1.4s + 0.806} \tag{42}$$

the lower boundary original model of tracking curve as

$$T_{RL}(j\omega) = \frac{0.9}{s + 0.9} \tag{43}$$

Adding two poles (p1=-1, p2=-4), which locate in left half s-plane to ensure stability of T_{RL} and are as close to the origin as possible but far enough away not to significantly affect the time response (see Sec.4.4.1), to eq. (43) to make lower tracking boundary separate from upper tracking boundary when upper tracking boundary cross over 0 dB line, then the final lower boundary transfer function is

$$T_{R_L}(j\omega) = \frac{3.6}{(s+0.9)(s+1)(s+4)} \tag{44}$$

Stability performance index and robust performance index are respectively

$$\left| \frac{q_{11}(s) g_1(s)}{1 + q_{11}(s) g_1(s)} \right| \leq 1.1$$

and

$$\left| \frac{q_{11}(s)}{1 + q_{11}(s) g_1(s)} \right| \leq 0.1$$

Corresponding minimum amplitude margin and phase margin are respectively

$$K_m = 1 + {}^{1}\!/_{u} = 1.9091 = 5.5155\text{dB}$$

and

$$\Phi_m = 180^\circ - \theta, \theta = \cos^{-1}(0.5 / \mu^2 - 1) = 54.062^\circ$$

2. **Plant Template and Border Calculation for Loop I.** According to the requirements of performance index, generate the tracking response boundary, robust stability boundary and inference rejection boundary in Nichols chart.
3. **Controller and Prefilter Design for Loop I.** In Fig. 21(a), the open-loop frequency characteristics curve (noted by black solid line) of the nominal plant (corresponding to G(s) =1) and the compound boundary (the region embraced by green and red solid line) are drawn up in Nichols chart. Apparently, the open-loop frequency curve locates under tracking performance boundary curve, open-loop frequency characteristics curve cross over the instability boundary (red solid ring line in Fig. 21(a)) which make the MISO system of loop I instable or unsatisfactory for corresponding performance requirements. So, it is necessary to enlarge the controller gain and introduce into dynamic compensation element to shape the open loop frequency characteristic curve to ensure shaped open-loop frequency characteristic meet the requirtments of stability and dynamic performance indics. Using MATLAB QFT toolbox, we get

$$g_1(s) = \frac{8.855(s/2.045+1)(s/8.68+1)}{(s/113.5+1)(s/907.9+1)} \tag{45}$$

$$f_{11}(s) = \frac{1.275}{s/0.6+1} \tag{46}$$

The open-loop frequency characteristics curve with G(s) is shown in Fig.21 (b). Clearly, the shaped curve does not cross over the instability region (red solid ring line),i.e. the shaped system is stable. Besides, the characteristic of tracking boundary is met.

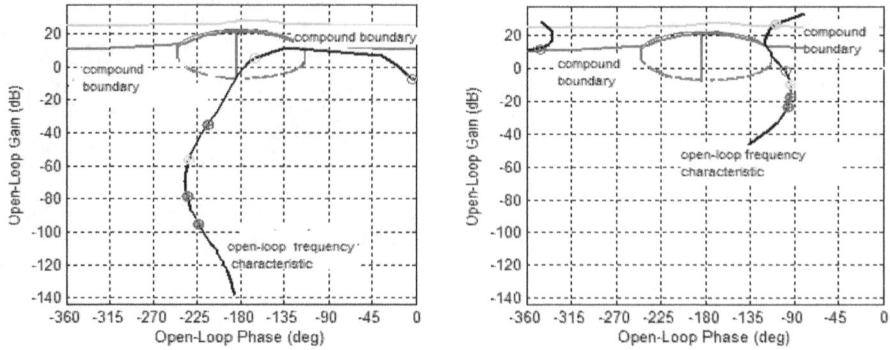

(a) Open-loop frequency response when G(s) =1 (b) Open-loop frequency response with controller

Fig. 21. Open loop frequency characteristics in Nichols Chart

4. **Verification and Simulation for Loop I.** Closed-loop system stability margin analysis curve, inference rejection boundary analysis curves and tracking boundary analysis curves in loop I are given in Fig.22 ,Fig.23 and Fig.24. Clearly, the stability margin curve, inference rejection boundary curve and tracking boundary curve are all under the stability performance index curve, the performance index curve and between the upper and lower boundaries of tracking curves. Obviously, Closed-loop control system satisfies the performance requirements in loop I.

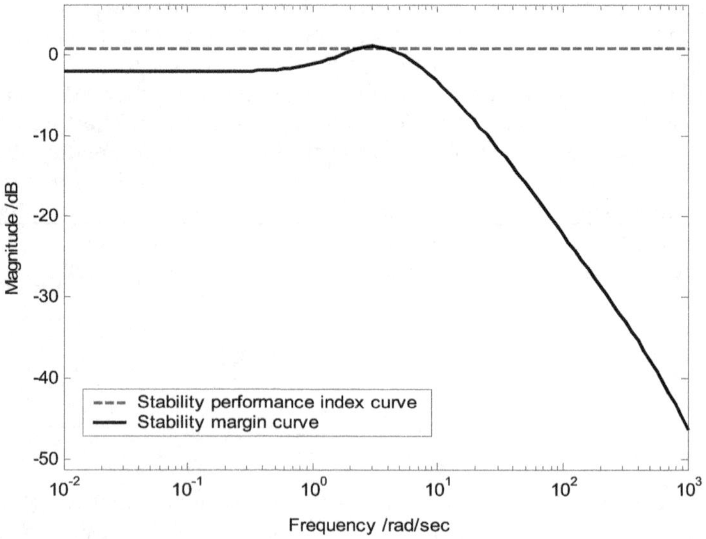

Fig. 22. Stability margin

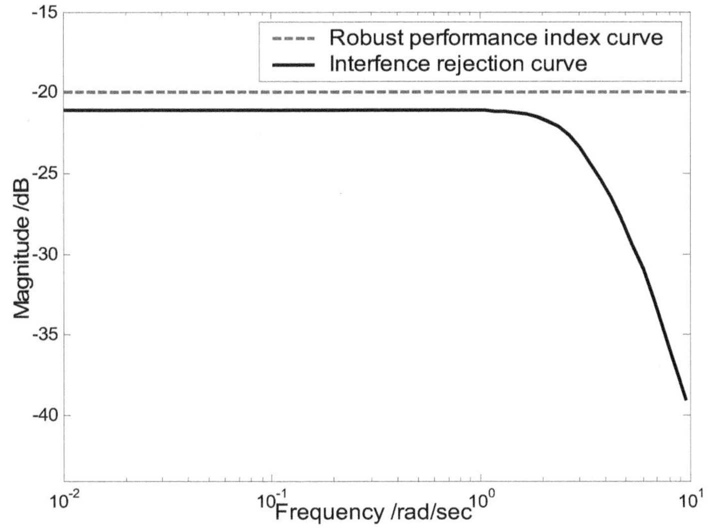

Fig. 23. Disturbance rejection boundary

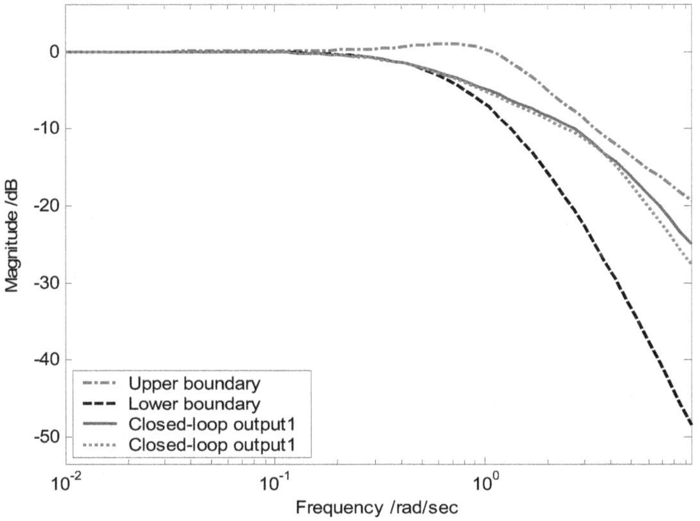

Fig. 24. Tracking boundary

The time-domain simulation results of closed-loop system under 5 design envelopes are shown in Fig.25 and Fig.26. The unit step-response of sideslip angle lies between the upper and lower boundary response curve; the unit step-response of disturbance input are located under the given boundary. Apparently, the closed-loop system satisfies the requirements of robust stability and tracking boundary requirements, and owns strong disturbance rejection capability.

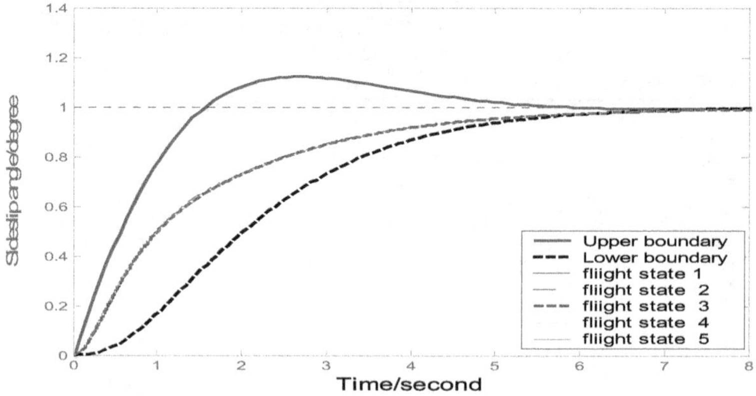

Fig. 25. The unit step response of β

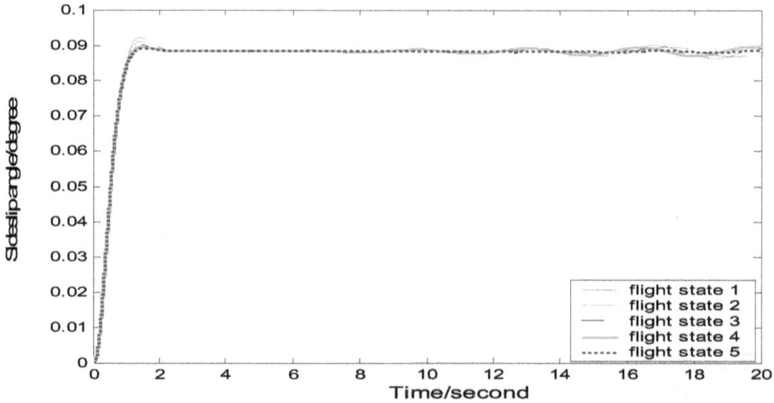

Fig. 26. The unit step response of β with disturbance

5.3.4 QFT design for loop II

QFT design for loop II is similar to that for loop I.

1. **Selection of Performance Indices.** Tracking performance indices of roll angle is overshoot $\sigma\% \le 5\%$ and settling time $t_s \le 12s$, the upper and lower boundary tracking curve are respectively

$$T_{RU}(j\omega) = \frac{0.25(1.7s+1)}{s^2+0.78s+0.25} \tag{47}$$

$$T_{RL}(j\omega) = \frac{3.6}{(s+0.9)(s+1)(s+4)} \tag{48}$$

Stability performance index and robust performance index are defined as

$$\left|\frac{q_{22}(s)g_2(s)}{1+q_{22}(s)g_2(s)}\right| \le \mu = 1.1 \text{ and } \left|\frac{q_{22}(s)}{1+q_{22}(s)g_2(s)}\right| \le 0.1$$

Minimum amplitude margin and phase margin are 5.5155B and 54.062° respectively.

2. **Controller and Prefilter Design for Loop II.**

Similar to loop I, using MATLAB QFT toolbox, we can get

$$g_2(s) = \frac{11.8(s/27.94+1)(s/1.18+1)}{(s/1280+1)(s/1926+1)} \tag{49}$$

$$f_{22}(s) = \frac{1.01}{(s/0.7+1)} \tag{50}$$

3. **Verification and Simulation for loop II.** Closed-loop system satisfies requirements of robust stability and tracking boundary requirements and owns strong disturbance rejection capability.

5.3.5 Performance analysis of QFT controller for the UAV's lateral motion

QFT control structure for the UAV's lateral motion is shown in Fig.27 .Given β_c and ϕ_c are 0, the initial value of ϕ is 5°, the initial sideslip angle β is 1°, substitute the UAV's lateral state equation, $g_1(s), f_{11}(s), g_2(s), f_{22}(s)$, models of rudder and ailerons into Fig.27. The simulation results are shown in Fig.28 and Fig.29. The overshoot of β is about 0.064° and settling time is about 1 second. The settling time of yaw angle rate, roll angle rate and roll angle are all about 0.1 second. Besides, the initial value of sideslip angle almost have no influence in roll angle response, the settling time of yaw angle rate, roll angle rate is no more than 1 second. Clearly, QFT controller for the UAV's lateral motion satisfies the requirements of performance indices, own better flight stability and robustness.

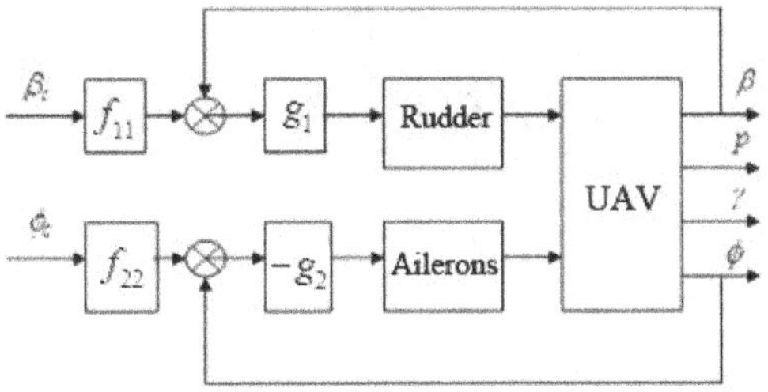

Fig. 27. QFT control structure for the UAV's lateral motion

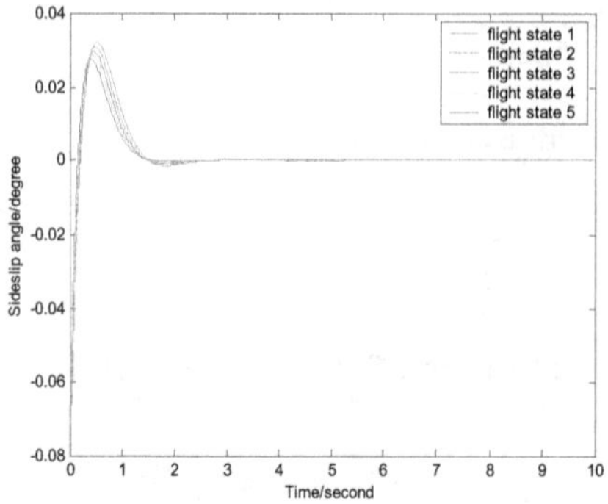

(a) Sideslip angle

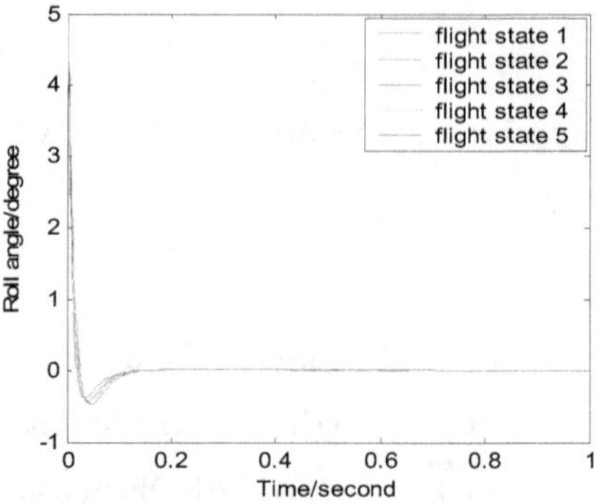

(b) Roll angle

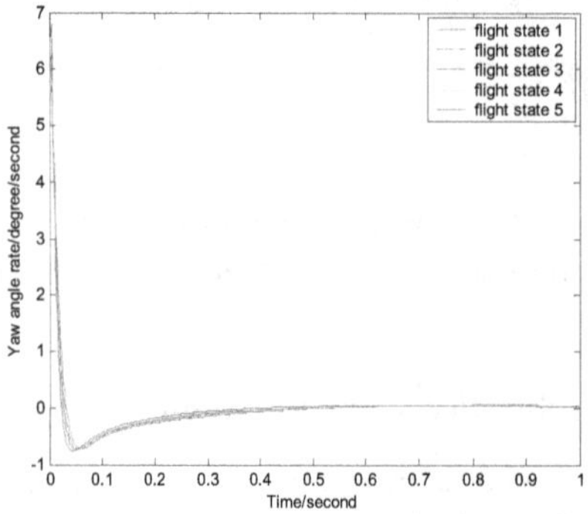

(c) Yaw angle rate

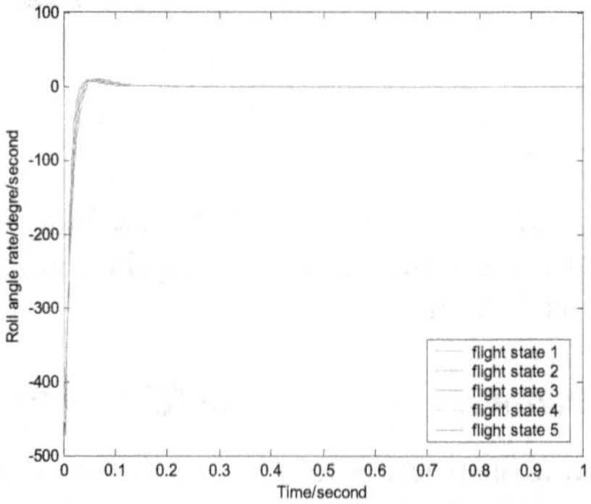

(d) Roll angle rate

Fig. 28. Responses of sideslip angle, roll angle, yaw angle rate and roll angle rate when $\phi_0 = 5^{\circ}$

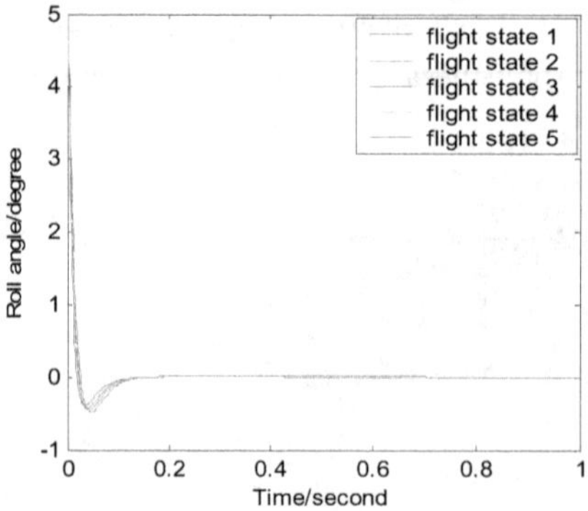

(a) Sideslip angle

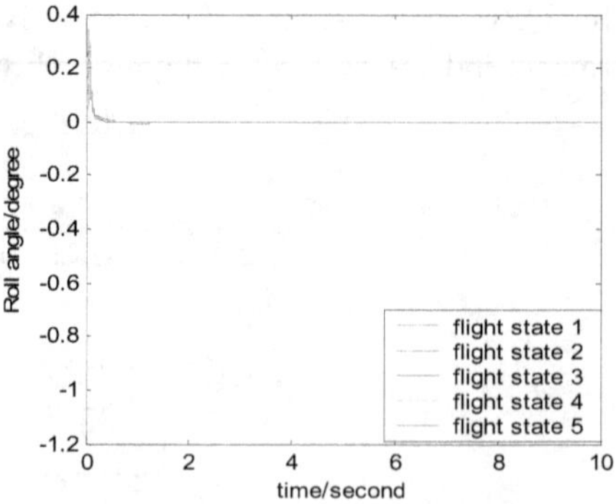

(b) Roll angle

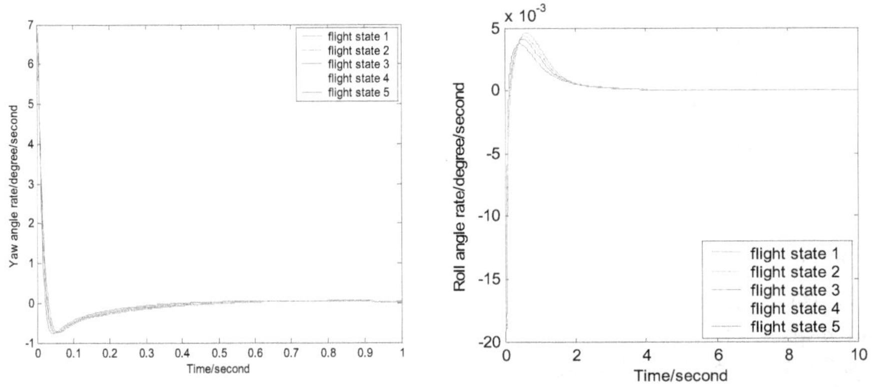

(c) Yaw angle rate (d) Roll angle rate

Fig. 29. Responses of sideslip angle, roll angle, yaw angle rate and roll angle rate when $\beta_0 = 1^\circ$

6. Summary

This chapter is devoted to presenting an overview and in-depth expression of QFT in order to enhance the understanding and appreciation of the power of the QFT technique. Then, A QFT design of robust controller for a certain UAV's lateral motion, which is a MIMO system, is proposed base on BNIA principle in order to show how to apply QFT in flight control system of UAVs. Meantime, the simulation results show that the QFT controller own better robust stability and superior dynamic characteristics which verify the validity of presented method.

7. Symbols & terminology

T_R Acceptable command or tracking input-output responses

$\Im_R$ A set of T_R

T_D Acceptable disturbance input-output responses

$\Im_D$ A set of T_D

P MISO plant with uncertainty

φ A set of P

MIMO Multiple-input multiple-output; more than one tracking and/or external disturbance inputs and more than one output

MISO Multiple-input single-output; a system having one tracking input, one or more external disturbance inputs, and a single output

$B_D(jw_i), B_R(jw_i), B_o(jw_i)$ The disturbance, tracking, and optimal bounds on $L(j\omega_i)$ for the MISO system

ω_h The frequency bandwidth

$\delta_i(j\omega)$ The magnitude variation due to the plant parameter uncertainty

Lm Log magnitude
LTI Linear-time-invariant
FOM figure of merit
ω_b The symbol for bandwidth frequency of the models
ω_m The resonant frequency
$\omega_\Phi, \omega_{\Phi i}$ Phase margin frequency for a MISO system and for the ith loop of a MIMO system, respectively
ω_s Sampling frequency
$R, R = \{r_i\}$ The tracking input for a MISO system and the tracking input vector for a MIMO system, respectively
$B_u = Lm\, T_{RU}$ The Lm of the desired tracking control ratio for the upper bound of the MISO system
$B_L = Lm\, T_{RL}$ The Lm of the desired tracking control ratio for the lower bound of the MISO system
B_s Stability bounds for the discrete design
$\delta_D(j\omega_i)$ The (upper) value of $Lm\, T_D(j\omega_i)$ for MISO system
$\delta_{hf}(j\omega_i)$ The dB difference between the augmented bounds of B_u and B_L in the high frequency range for a MISO system
$\delta_R(j\omega_i)$ The dB difference between B_u and B_L for a given frequency ω, for a MISO system
$F, F = \{f_{ij}\}$ The prefilter for a MISO system and the mxm prefilter matrix for a MIMO system respectively
$G, G = \{f_{ij}\}$ The compensator or controller for a MISO system and the mxm compensator or controller matrix for a MIMO system, respectively. For a diagonal matrix $G = \{f_{ij}\}$
γ, γ_i The phase margin angle for the MISO system and for the ith loop of the MIMO system, respectively
J The number of plant transfer functions for a MISO system or plant matrix for a MIMO system that describes the region of plant parameter uncertainty where i = 1, 2.....J denotes the particular plant case in the region of plant parameter uncertainty
λ The excess of poles over zeros of a transfer function
L_o, L_{oi} The optimal loop transmission function for the MISO system and the ith loop of the MIMO system, respectively
M_L, M_u The specified closed-loop frequency domain overshoots constraint for the MISO system and for the ith loop of a MIMO system, respectively. This overshoot constraint may be dictated by the phase margin angle for the specified loop transmission function
$\Im P(j\omega_i)$ Script cap tee in conjunction with P denotes a template, i.e., $\Im P(j\omega_i)$ and $\Im Q(j\omega_i)$ frequency, for a MISO and MIMO plants respectively
T_{RU} The desired MISO tracking control ratio that satisfies the specified upper bound FOM

T_{R_L} The desired MISO tracking control ratio that satisfies the specified lower bound FOM
T_D The desired MISO disturbance control ratio which satisfies the specified FOM
UAV Unmanned Aerial Vehicle
BNIA Basically Non-interacting

8. Acknowledgement

The work of this chapter is supported by Natural Science Basic Research Plan in Shaanxi Province of China (Program No. 2011GQ8005) and Northwestern Polytechnical University Foundation for Fundamental Research (No. :NPU-FFR-JC20100216)

9. References

Chen Huaimin: An Integrated QFT/EA Controller Design Method for a UAV's Lateral Flight Control System, Mechanical Science and Technology, Vol.27-3(2008),p. 413.

Constantine H. Houpis,Steven J. Rasmussen. Quantitative Feedback Theory: Fundamentals and Applications[M], Marcel Dekker, Inc. New York, Basel

Horowitz I. M, and M. Sidi, "Synthesis of Feedback Systems with Large Plant Ignorance for Prescribed Time Domain Tolerances," Int. J. of Control, vol. 16, pp 287-309, 1973.

Horowitz, I. M. and C. Loecher, "Design 3x3 Multivariable Feedback System with Large Plant Uncertainty," hit. J. Control, vol. 33, pp. 677-699,1981.

Ibid, "Synthesis of Feedback Systems with Non-Linear Time Uncertain Plants to Satisfy Quantitative Performance Specifications ," IEEE Proc., vol. 64, pp. 123-130,1976.

Houpis, C. H. "Quantitative Feedback Theory (QFT) for the Engineer: A Paradigm for the Design of Control Systems for Uncertain Systems," WL-TR-95-3061, AF Wright Laboratory, Wright-Patterson AFB, OH, 1995 (Available from National Technical Information Service, 5285 Port Royal Road, Springfield, VA 22151, document number AD-A297571.)

Houpis, C. H. and P. R. Chandler, Editors: "Quantitative Feedback Theory Symposium Proceedings," WL-TR-92-3063, Wright Laboratories, Wright-Patterson AFB,OH, 1992.

O Yaniv,Y Chait:A Simplified Multi-Input Multi-Output Formulation for Quantitative Feedback Theory, Journal of Dynamic Systems, Measurement, and Control,Vol.114-6(1998),p.179.

Reynolds, O. R., M Pachter, and C. H. Houpis. "Design of a Subsonic Flight Control System for the Vista F-16 Using Quantitative Feedback Theory, "Proceedings of the American Control Conference, pp. 350-354,1994.

Thompson, D. F., and O. D. I. Nwokah, "Optimal Loop Synthesis in Quantitative Feedback Theory," Proceedings, of the American Control Conference, San Diego, CA, pp.626-631,1990.

Trosen, D. W., M, Pachter, and C. H. Houpis, "Formation Flight Control Automation," Proceedings of the American Institute of Aeronautics and Astronautics (AIAA) Conference, pp. 1379-1404, Scottsdale, AZ, 1994.

Part 2

Adaptive and Fault Tolerant Flight Control

4

Fault Tolerant Flight Control Techniques with Application to a Quadrotor UAV Testbed

Youmin Zhang and Abbas Chamseddine
Department of Mechanical and Industrial Engineering, Concordia University
Canada

1. Introduction

Unmanned Aerial Vehicles (UAVs) are gaining more and more attention during the last few years due to their important contributions and cost-effective applications in several tasks such as surveillance, search and rescue missions, geographic studies, as well as various military and security applications. Due to the requirements of autonomous flight under different flight conditions without a pilot onboard, control of UAV flight is much more challenging compared with manned aerial vehicles since all operations have to be carried out by the automated flight control, navigation and guidance algorithms embedded on the onboard flight microcomputer/microcontroller or with limited interference by a ground pilot if needed.

As an example of UAV systems, the quadrotor helicopter is relatively a simple, affordable and easy to fly system and thus it has been widely used to develop, implement and test-fly methods in control, fault diagnosis, fault tolerant control as well as multi-agent based technologies in formation flight, cooperative control, distributed control, mobile wireless networks and communications. Some theoretical works consider the problems of control (Dierks & Jagannathan, 2008), formation flight (Dierks & Jagannathan, 2009) and fault diagnosis (Nguyen et al., 2009; Rafaralahy et al., 2008) of the quadrotor UAV. However, few research laboratories are carrying out advanced theoretical and experimental works on the system. Among others, one may cite for example, the UAV SWARM health management project of the Aerospace Controls Laboratory at MIT (SWARM, 2011), the Stanford Testbed of Autonomous Rotorcraft for Multi-Agent Control (STARMAC) project (STARMAC, 2011) and the Micro Autonomous Systems Technologies (MAST) project (MAST, 2011).

A team of researchers at the Department of Mechanical and Industrial Engineering of Concordia University, with the financial support from NSERC (Natural Sciences and Engineering Research Council of Canada) through a Strategic Project Grant and a Discovery Project Grant and three Canadian-based industrial partners (Quanser Inc., Opal-RT Technologies Inc., and Numerica Technologies Inc.) as well as Defence Research and Development Canada (DRDC) and Laval University, have been working on a research and development project on fault-tolerant and cooperative control of multiple UAVs since 2007 (see the Networked Autonomous Vehicles laboratory (NAV, 2011)). In addition to the work that has been carried out for the multi-vehicles case, many fault-tolerant control (FTC) strategies have been developed and applied to the single vehicle quadrotor helicopter UAV system. The objective is to consider actuator faults and to propose FTC methods to accommodate as much as possible the fault effects on the system performance. The proposed methods have been tested either in simulation, experimental or both frameworks where the experimental implementation has been carried out using a cutting-edge quadrotor UAV testbed known also as Qball-X4. The developed approaches include the Gain-Scheduled PID (GS-PID), Model Reference Adaptive Control (MRAC), Sliding Mode Control (SMC), Backstepping Control (BSC), Model Predictive Control (MPC) and Flatness-based Trajectory Planning/Re-planning (FTPR), etc. Some of these methods require an information about the time of occurrence, the location and the amplitude of faults whereas others do not. In the former case, a fault detection and diagnosis (FDD) module is needed to detect, isolate and identify the faults which occurred.

Table 1 summarizes the main fault-tolerant control methodologies that have been developed and applied to the quadrotor helicopter UAV. The table shows which methods require an FDD scheme and whether they have been tested in simulation or experimentally. In the case of experimental application, one can also distinguish between two categories of faults: a simulated fault and a real damage. In the first case, a fault has been generated in an actuator by multiplying its control input by a gain smaller than one, thus simulating a loss in the control effectiveness. In the second case, the fault/damage has been introduced by breaking the tip of a propeller of the Qball-X4 UAV during flight.

Strategy	Passive/Active	Need for FDD	Simulation	Experiment	Real damage
GS-PID	Active	✓	✓	✓	X
MRAC	Active	X	X	✓	✓
SMC	Passive	X	✓	✓	✓
BSC	Passive	X	✓	X	X
MPC	Active	✓	✓	X	X
FTPR	Active	✓	✓	✓	X

Table 1. Fault-tolerant control methods (Note: (✓) Yes/Done; (X) No/Not done yet).

These methods will be discussed in the subsequent sections. The remainder of this chapter is then organized as follows. Section 2 introduces the quadrotor UAV system that is used as a testbed for the proposed methods. Section 3 presents the FTC methods summarized in Table 1. The simulation as well as the experimental flight results are given and discussed in Section 4. Some concluding remarks together with future work are finally given. Note that due to space consideration, only some of the developed control methods will be included in this chapter. Interested readers can refer to the website of the NAV Lab on '`http://users.encs.concordia.ca/~ymzhang/UAVs.htm`' for more information.

2. Description and dynamics of the quadrotor UAV system

The quadrotor UAV of the Department of Mechanical and Industrial Engineering of Concordia University is the Qball-X4 testbed (Figure 1(a)) developed by Quanser Consulting Inc. The quadrotor UAV is enclosed within a protective carbon fiber round cage (therefore a name of Qball-X4) to ensure safe operation of the vehicle and protection to the personnel who is working with the vehicle in an indoor research and development environment. It uses four 10-inch propellers and standard RC motors and speed controllers. It is equipped with the Quanser Embedded Control Module (QECM), which is comprised of a Quanser HiQ aero data acquisition card and a QuaRC-powered Gumstix embedded computer where QuaRC is Quanser's Real-time Control software. The Quanser HiQ provides high-resolution accelerometer, gyroscope, and magnetometer IMU sensors as well as servo outputs to drive four motors. The on-board Gumstix computer runs QuaRC, which allows to rapidly develop and deploy controllers designed in MATLAB/Simulink environment to real-time control the Qball-X4. The controllers run on-board the vehicle itself and runtime sensors measurement, data logging and parameter tuning is supported between the host PC and the target vehicle (Quanser, 2010).

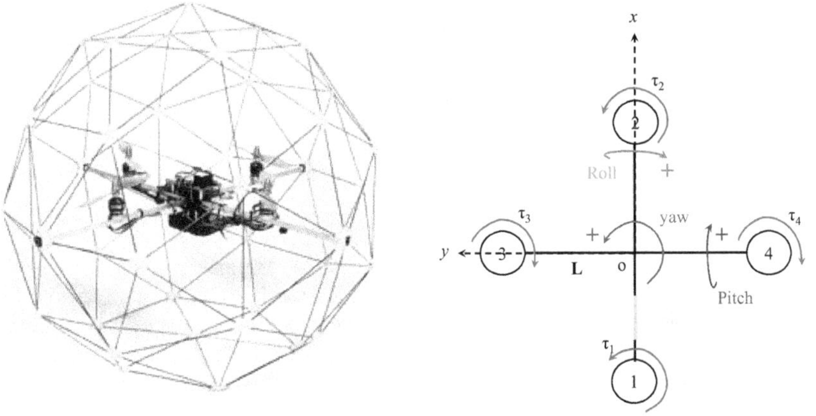

(a) The Quanser Qball-X4 quadrotor UAV (b) Qball-X4 schematic representation

Fig. 1. The Quanser Qball-X4 quadrotor UAV and its schematic representation.

The block diagram of the entire UAV system is illustrated in Figure 2. It is composed of three main parts:

- The first part represents the Electronic Speed Controllers (ESCs) + the motors + the propellers in a set of four. The input to this part is $u = [u_1\ u_2\ u_3\ u_4]^T$ which are Pulse Width Modulation (PWM) signals. The output is the thrust vector $T = [T_1\ T_2\ T_3\ T_4]^T$ generated by four individually-controlled motor-driven propellers.
- The second part is the geometry that relates the generated thrusts to the applied lift and torques to the system. This geometry corresponds to the position and orientation of the propellers with respect to the system's center of mass.
- The third part is the dynamics that relate the applied lift and torques to the position (P), velocity (V) and acceleration (A) of the Qball-X4.

The subsequent sections describe the corresponding mathematical model for each of the blocks of Figure 2.

2.1 Qball-X4 dynamics

The Qball-X4 dynamics in a hybrid coordinate system are given hereafter where the position dynamics are expressed in the inertial frame and the angular dynamics are expressed in the

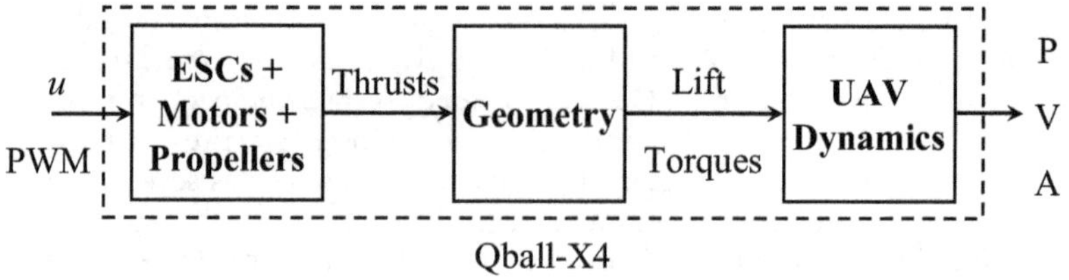

Fig. 2. The UAV system block diagram.

body frame (Bresciani, 2008):

$$\begin{aligned}
m\ddot{x} &= u_z\,(cos\phi\; sin\theta\; cos\psi + sin\phi\; sin\psi) - k_x\dot{x}\\
m\ddot{y} &= u_z\,(cos\phi\; sin\theta\; sin\psi - sin\phi\; cos\psi) - k_y\dot{y}\\
m\ddot{z} &= u_z\,(cos\phi\; cos\theta) - mg - k_z\dot{z}\\
J_x\dot{p} &= u_p + (J_y - J_z)\,qr - J_T q\,\Omega - k_p p\\
J_y\dot{q} &= u_q + (J_z - J_x)\,pr - J_T p\,\Omega - k_q q\\
J_z\dot{r} &= u_r + (J_x - J_y)\,pq - k_r r
\end{aligned} \tag{1}$$

where x, y and z are the coordinates of the quadrotor UAV center of mass in the inertial frame. m is the system mass and J_x, J_y and J_z are the moments of inertia along y, x and z directions respectively. θ, ϕ and ψ are the pitch, roll and yaw Euler angles and p, q and r are the angular velocities in the body-fixed frame. k_x, k_y, k_z, k_p, k_q and k_r are drag coefficients and are constant. J_T is the moment of inertia for each motor and Ω is the overall speed of propellers:

$$\Omega = -\Omega_1 - \Omega_2 + \Omega_3 + \Omega_4 \tag{2}$$

where Ω_i is the ith propeller speed.

The angular velocities in the inertial frame (Euler rates) can be related to those in the body frame as follows:

$$\begin{bmatrix} p\\ q\\ r \end{bmatrix} = \begin{bmatrix} 1 & 0 & -sin\theta\\ 0 & cos\phi & cos\theta sin\phi\\ 0 & -sin\phi & cos\theta cos\phi \end{bmatrix} = \begin{bmatrix} \dot{\phi}\\ \dot{\theta}\\ \dot{\psi} \end{bmatrix} \tag{3}$$

Close to hovering conditions, the matrix in the above equation is close to identity matrix and therefore the angular velocities in the body frame can be seen as the angular velocities in the inertial frame. The model (1) can then be written as:

$$\begin{aligned}
m\ddot{x} &= u_z\,(cos\phi\; sin\theta\; cos\psi + sin\phi\; sin\psi) - k_x\dot{x}\\
m\ddot{y} &= u_z\,(cos\phi\; sin\theta\; sin\psi - sin\phi\; cos\psi) - k_y\dot{y}\\
m\ddot{z} &= u_z\,(cos\phi\; cos\theta) - mg - k_z\dot{z}\\
J_x\ddot{\theta} &= u_\theta + (J_y - J_z)\,\dot{\phi}\dot{\psi} - J_T\dot{\phi}\,\Omega - k_\theta\dot{\theta}\\
J_y\ddot{\phi} &= u_\phi + (J_z - J_x)\,\dot{\theta}\dot{\psi} - J_T\dot{\theta}\,\Omega - k_\phi\dot{\phi}\\
J_z\ddot{\psi} &= u_\psi + (J_x - J_y)\,\dot{\theta}\dot{\phi} - k_\psi\dot{\psi}
\end{aligned} \tag{4}$$

where u_p, u_q, u_r, k_p, k_q and k_r have been respectively changed to u_θ, u_ϕ, u_ψ, k_θ, k_ϕ, k_ψ for notation convenience. At low speeds, one can obtain a simplified nonlinear model of (4) by neglecting drag terms and gyroscopic and Coriolis-centripetal effects:

$$\begin{aligned} m\ddot{x} &= u_z\,(cos\phi\; sin\theta\; cos\psi + sin\phi\; sin\psi) \\ m\ddot{y} &= u_z\,(cos\phi\; sin\theta\; sin\psi - sin\phi\; cos\psi) \\ m\ddot{z} &= u_z\,(cos\phi\; cos\theta) - mg \\ J_x\ddot{\theta} &= u_\theta \\ J_y\ddot{\phi} &= u_\phi \\ J_z\ddot{\psi} &= u_\psi \end{aligned} \tag{5}$$

A further simplified linear model can be obtained by assuming hovering conditions ($u_z \approx mg$ in the x and y directions) with no yawing ($\psi = 0$) and small roll and pitch angles as follows:

$$\begin{aligned} \ddot{x} &= \theta g; & J_x\ddot{\theta} &= u_\theta \\ \ddot{y} &= -\phi g; & J_y\ddot{\phi} &= u_\phi \\ \ddot{z} &= u_z/m - g; & J_z\ddot{\psi} &= u_\psi \end{aligned} \tag{6}$$

The nonlinear model (4) will be used later on in the design of fault-tolerant control strategies such as the BSC, the MPC and the SMC. The simplified model (6) will be used for the MRAC design as well as for the trajectory planning and re-planning approach.

2.2 ESCs, motors and propellers

The motors of the Qball-X4 are outrunner brushless motors. The generated thrust T_i of the ith motor is related to the ith PWM input u_i by a first-order linear transfer function:

$$T_i = K\frac{\omega}{s+\omega}u_i \;\; ; \;\; i = 1, ..., 4 \tag{7}$$

where K is a positive gain and ω is the motor bandwidth. K and ω are theoretically the same for the four motors but this may not be the case in practice and therefore, this can be one of sources of modeling errors/uncertainties for fault-tolerant control design and trajectory planning/re-planning.

2.3 Geometry

A schematic representation of the Qball-X4 is given in Figure 1(b). The motors and propellers are configured in such a way that the back and front (1 and 2) motors spin clockwise and the left and right (3 and 4) spin counter-clockwise. Each motor is located at a distance L from the center of mass o and when spinning, a motor produces a torque τ_i. The origin of the body-fixed frame is the system's center of mass o with the x-axis pointing from back to front and the y-axis pointing from right to left. The thrust T_i generated by the ith propeller is always pointing upward in the z-direction in parallel to the motor's rotation axis. The thrusts T_i and the torques τ_i result in a lift in the z-direction (body-fixed frame) and torques about the x, y and z axis. The relation between the lift/torques and the thrusts is:

$$\begin{aligned} u_z &= T_1 + T_2 + T_3 + T_4 \\ u_\theta &= L(T_1 - T_2) \\ u_\phi &= L(T_3 - T_4) \\ u_\psi &= \tau_1 + \tau_2 - \tau_3 - \tau_4 \end{aligned} \tag{8}$$

The torque τ_i produced by the ith motor is directly related to the thrust T_i via the relation of $\tau_i = K_\psi T_i$ with K_ψ as a constant. In addition, by setting $T_i \approx K u_i$ from (7), the relation (8) reads:

$$\begin{aligned} u_z &= K(u_1 + u_2 + u_3 + u_4) \\ u_\theta &= KL(u_1 - u_2) \\ u_\phi &= KL(u_3 - u_4) \\ u_\psi &= KK_\psi(u_1 + u_2 - u_3 - u_4) \end{aligned} \tag{9}$$

where u_z is the total lift generated by the four propellers and applied to the quadrotor UAV in the z-direction (body-fixed frame). u_θ, u_ϕ and u_ψ are respectively the applied torques in θ, ϕ and ψ directions (see Figure 1(b)).

3. Fault-tolerant control algorithms

Modern technological systems rely on sophisticated control systems to replace or reduce the intervention of human operators. In the event of malfunctions in actuators, sensors or other system components, a conventional feedback control design may result in an unsatisfactory performance or even instability of the controlled system. To prevent such situations, new control approaches have been developed in order to tolerate component malfunctions while maintaining desirable stability and performance properties. This is particularly important for safety-critical systems, such as aircrafts, spacecrafts, nuclear power plants, and chemical plants processing hazardous materials. In such systems, the consequences of a minor fault in a system component can be catastrophic. It is necessary then to design control systems which are capable of tolerating potential faults in these systems in order to improve the reliability and availability while providing a desirable performance. These types of control systems are often known as fault-tolerant control systems (FTCS). More precisely, FTCS are control systems which possess the ability to accommodate component failures automatically. They are capable of maintaining overall system stability and acceptable performance in the event of such failures (Zhang & Jiang, 2008).

Generally speaking, FTCS can be classified into two types: passive (PFTCS) and active (AFTCS). In PFTCS, controllers are fixed and are designed to be robust against a class of presumed faults. This approach needs neither FDD schemes nor controller reconfiguration, but it has limited fault-tolerant capabilities. In contrast to PFTCS, AFTCS react to the system component failures actively by reconfiguring control actions so that the stability and acceptable performance of the entire system can be maintained (Zhang & Jiang, 2008). An overall structure of a typical AFTCS is shown in Figure 3. In the FDD module, faults must be detected and isolated as quickly as possible, and fault parameters, system state/output variables, and post-fault system models need to be estimated on-line in real-time. Based on the on-line information on the post-fault system model, the controller must be automatically reconfigured to maintain stability, desired dynamic performance and steady-state performance. To avoid potential actuator saturation and to take into consideration the degraded performance after fault occurrence, a command/reference governor may also need to be designed to adjust command input or reference trajectory automatically. Interested readers about FTCS may refer to the bibliographical review (Zhang & Jiang, 2008) and the recently published books (Noura et al., 2009) and (Edwards et al., 2010).

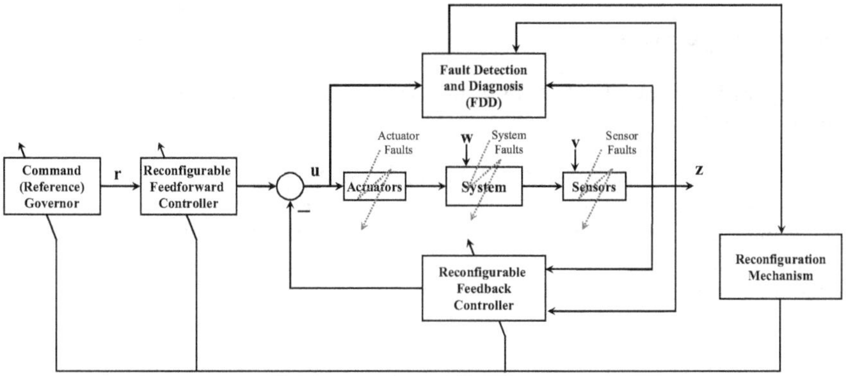

Fig. 3. A general structure of AFTCS (Zhang & Jiang, 2008).

The subsequent sections consider some of the FTC methods that have been developed for the quadrotor UAV. Some of these are classified as passive FTC methods where the fault tolerance is achieved thanks to the controller's robustness without changing the controller gains. Others are active FTC methods where control gains are updated in function of the occurring faults.

3.1 Gain-scheduled PID (GS-PID)

The Proportional-Integral-Derivative (PID) controller is the most widely used among controllers in industry. This is due to its unique features such as the simple structure, the ease of use and tunning. Its main advantages over other control strategies is that it does not require a mathematical model of the process/system to be controlled and thus allows to save time and effort by skipping the modeling phase. PID controllers are reliable and can be used for linear and nonlinear systems with certain level of robustness to model uncertainties and disturbances. Although one single PID controller can handle a wide range of system nonlinearities, better performance can be obtained when using multiple PIDs to cover the entire operation range of a nonlinear process/system. This is known as gain-scheduled PID (GS-PID) (Sadeghzadeh et al., 2011).

The operating principle of GS-PID is shown in Figure 4(a) where the controlled system may have varying dynamic properties as for example a varying gain (Haugen, 2004). The adjustment/scheduling of the PID controller gains is performed by using a gain scheduling variable GS. This is some measured process variable which at every instant of time expresses or represents the dynamic properties of the process.

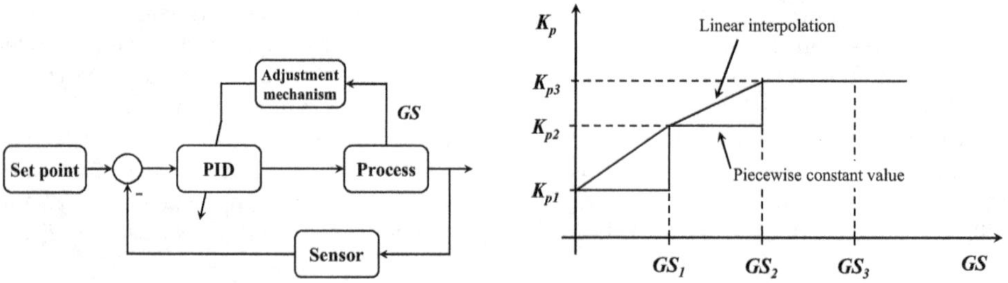

(a) Operating principle of GS-PID (b) Constant and linear interpolation

Fig. 4. The operating principle of GS-PID and two rules for the interpolation of the proportional gain K_p.

There are several ways to express the PID parameters as functions of the GS variable such as the *piecewise constant controller parameters* and the *piecewise interpolation*. In the former method, an interval is defined around each GS value and the controller parameters are kept constant as long as the GS value is within the interval (see Figure 4(b) for an example of the proportional gain K_p). When the GS variable changes from one interval to another, the controller parameters are changed abruptly (Haugen, 2004). The same idea applies for the Integrator and Derivative gains. In the second method also shown in Figure 4(b), a linear function is found relating the controller parameter (output variable) and the GS variable (input variable). For the proportional gain, the linear function is of the form $K_p = aGS + b$ where a and b are two constants to be calculated.

Since faults can be seen as varying parameters, GS-PID can also be used to deal with possible fault conditions that may take place in the actuators or the system components. Some research works consider this problem: a GS-PID control strategy is proposed in (Bani Milhim, 2010a) and (Bani Milhim et al., 2010b) in simulation framework to achieve fault-tolerant control for the quadrotor helicopter UAV in the presence of actuator faults. In (Johnson et al., 2010), the authors investigate this technique for a Georgia Tech Twinstar fixed-wing research vehicle in the presence of partial damage of the wing. As continuation of the work presented in (Bani Milhim et al., 2010b), GS-PID has been considered for further investigation and most importantly for experimental implementation and application to the Qball-X4 UAV testbed for fault-tolerant trajectory tracking control (Sadeghzadeh et al., 2011). The GS-PID has been implemented for different sections of the entire flight envelope by properly tuning the PID controller gains for both normal and fault conditions. A set of PID controllers is then designed for the fault-free situation and each possible fault situation. The switching from one PID to another is then based on the actuator's health status (hence, the scheduling variable GS is the actuator's health status). A Fault Detection and Diagnosis (FDD) scheme is then needed to provide the time of fault occurrence as well as the location and the magnitude of the fault during the flight. Based on the information provided by the FDD module, the GS-PID controller will switch the controller gains under normal flight conditions to the pre-tuned and fault-related gains to handle the faults during the flight of the UAV. One of the main issues to consider in GS-PID is how fast to switch from the nominal PID gains to the pre-tuned fault-related gains after fault occurrence. This issue does not represent a problem when the scheduling variable GS is a measured variable since in most cases it is instantaneously provided by the sensors. However, when the GS is the health status of the actuators then clearly, the switching time and gains depend on how fast and precise the FDD module is in detecting, isolating and identifying the faults. It is shown in Section 4 and through experimentations how the switching time affects the behavior of the system in handling the occurring faults.

3.2 Model reference adaptive control (MRAC)

Many adaptive control techniques for preserving stability have been proposed to deal with disturbances, unmodeled dynamics and time delays. Particularly, the concept of model reference adaptive control (MRAC) has gained significant attention and is now part of many standard textbooks on nonlinear and adaptive control. Two basic approaches for MRAC can be distinguished: the direct and the indirect approaches. In the direct method, the controller parameters are adjusted based on the error between the reference model describing the desired dynamics and the closed-loop dynamics of the physical plant. In the indirect approach, the parameters of the plant are estimated by updating them based on the identification error between the measured states and those provided by the estimation model. In this work, three MRAC techniques are implemented and applied to the Qball-X4, namely the MIT rule-based MRAC, the conventional MRAC and the modified MRAC (Chamseddine et al., 2011a).

3.2.1 MIT rule

This MRAC approach has been developed around 1960 at the Massachusetts Institute of Technology (MIT) for aerospace applications (Ioannou & Sun, 1995). For illustration, consider the plant:

$$\ddot{y}(t) = -a_1\dot{y}(t) - a_2y(t) + bu(t) \tag{10}$$

where a_1, a_2 and b are the unknown plant parameters. The reference model to be matched by the closed-loop plant is:

$$y_m^{(3)}(t) = -a_{m_1}\ddot{y}_m(t) - a_{m_2}\dot{y}_m(t) - a_{m_3}y_m(t) + b_m r(t) \tag{11}$$

where $r(t)$ is the reference command and a_{m_i} $(i = 1, 2, 3)$ and b_m are constant and are chosen according to performance specifications. Let the control input $u(t)$ be defined as follows:

$$u(t) = -k_1\dot{y}(t) - k_2y(t) - k_3\int_0^t (y(\tau) - r(\tau))\,d\tau \tag{12}$$

By replacing (12) in (10) and differentiating with respect to time, one obtains:

$$y^{(3)}(t) = -(a_1 + bk_1)\,\ddot{y}(t) - (a_2 + bk_2)\,\dot{y}(t) - bk_3y(t) + bk_3r(t) \tag{13}$$

It is obvious that one can achieve perfect model following if k_1, k_2 and k_3 are chosen such that:

$$a_1 + bk_1 = a_{m_1},\ a_2 + bk_2 = a_{m_2} \text{ and } bk_3 = a_{m_3} = b_m \tag{14}$$

The control signal given by (12) cannot be implemented since the system parameters a_1, a_2 and b are assumed to be unknown. Nevertheless, one can use the following:

$$u(t) = -\hat{k}_1(t)\dot{y}(t) - \hat{k}_2(t)y(t) - \hat{k}_3(t)\int_0^t (y(\tau) - r(\tau))\,d\tau \tag{15}$$

where $\hat{k}_i$ are the estimates of k_i and are updated according to the MIT rule. The objective of the MIT rule is to adjust the parameters k_1, k_2 and k_3 so as to minimize a cost function J. This cost function can be chosen for example as follows:

$$J = \frac{1}{2}e^2 \tag{16}$$

where e is the tracking error between the system and the reference model, i.e. $e = y - y_m$. It is reasonable to adjust the parameters in the direction of the negative gradient of J:

$$\frac{d\hat{k}_i}{dt} = -\gamma\frac{\partial J}{\partial \hat{k}_i} = -\gamma e\frac{\partial e}{\partial \hat{k}_i} = -\gamma e\frac{\partial y}{\partial \hat{k}_i} \tag{17}$$

where $\gamma > 0$ is the adaptation rate. $\partial e/\partial \hat{k}_i$ is called the sensitivity derivative of the system and is evaluated under the assumption that $\hat{k}_i$ varies slowly. To calculate $\partial y/\partial \hat{k}_i$ in (17) one can use (13) to obtain:

$$\begin{aligned}
\frac{\partial y^{(3)}}{\partial \hat{k}_1} &= -a_1\frac{\partial \ddot{y}}{\partial \hat{k}_1} - a_2\frac{\partial \dot{y}}{\partial \hat{k}_1} - b\hat{k}_3\frac{\partial y}{\partial \hat{k}_1} - b\ddot{y}(t) - b\hat{k}_1\frac{\partial \ddot{y}}{\partial \hat{k}_1} - b\hat{k}_2\frac{\partial \dot{y}}{\partial \hat{k}_1} \\
\frac{\partial y^{(3)}}{\partial \hat{k}_2} &= -a_1\frac{\partial \ddot{y}}{\partial \hat{k}_2} - a_2\frac{\partial \dot{y}}{\partial \hat{k}_2} - b\hat{k}_3\frac{\partial y}{\partial \hat{k}_2} - b\dot{y}(t) - b\hat{k}_1\frac{\partial \ddot{y}}{\partial \hat{k}_2} - b\hat{k}_2\frac{\partial \dot{y}}{\partial \hat{k}_2} \\
\frac{\partial y^{(3)}}{\partial \hat{k}_3} &= -a_1\frac{\partial \ddot{y}}{\partial \hat{k}_3} - a_2\frac{\partial \dot{y}}{\partial \hat{k}_3} - b\hat{k}_3\frac{\partial y}{\partial \hat{k}_3} - by(t) - b\hat{k}_1\frac{\partial \ddot{y}}{\partial \hat{k}_3} - b\hat{k}_2\frac{\partial \dot{y}}{\partial \hat{k}_3} + br(t)
\end{aligned} \tag{18}$$

With the assumption that the rate of adaptation is slow, i.e. $\dot{\hat{k}}_i$ are small and the changes of $y^{(3)}$, $\ddot{y}$ and y with respect to $\hat{k}_1$, $\hat{k}_2$ and $\hat{k}_3$ are also small, one can interchange the order of differentiation to obtain:

$$\begin{aligned}
\frac{d^3}{dt^3}\frac{\partial y}{\partial \hat{k}_1} &= -(a_1 + b\hat{k}_1)\frac{d^2}{dt^2}\frac{\partial y}{\partial \hat{k}_1} - (a_2 + b\hat{k}_2)\frac{d}{dt}\frac{\partial y}{\partial \hat{k}_1} - b\hat{k}_3\frac{\partial y}{\partial \hat{k}_1} - b\ddot{y}(t) \\
\frac{d^3}{dt^3}\frac{\partial y}{\partial \hat{k}_2} &= -(a_1 + b\hat{k}_1)\frac{d^2}{dt^2}\frac{\partial y}{\partial \hat{k}_2} - (a_2 + b\hat{k}_2)\frac{d}{dt}\frac{\partial y}{\partial \hat{k}_2} - b\hat{k}_3\frac{\partial y}{\partial \hat{k}_2} - b\dot{y}(t) \\
\frac{d^3}{dt^3}\frac{\partial y}{\partial \hat{k}_3} &= -(a_1 + b\hat{k}_1)\frac{d^2}{dt^2}\frac{\partial y}{\partial \hat{k}_3} - (a_2 + b\hat{k}_2)\frac{d}{dt}\frac{\partial y}{\partial \hat{k}_3} - b\hat{k}_3\frac{\partial y}{\partial \hat{k}_3} - by(t) + br(t)
\end{aligned} \tag{19}$$

These latter equations can be written as:

$$\begin{aligned}
\frac{\partial y}{\partial \hat{k}_1} &= \frac{-b}{p^3 + (a_1 + b\hat{k}_1)p^2 + (a_2 + b\hat{k}_2)p + b\hat{k}_3}\ddot{y}(t) \\
\frac{\partial y}{\partial \hat{k}_2} &= \frac{-b}{p^3 + (a_1 + b\hat{k}_1)p^2 + (a_2 + b\hat{k}_2)p + b\hat{k}_3}\dot{y}(t) \\
\frac{\partial y}{\partial \hat{k}_3} &= \frac{-b}{p^3 + (a_1 + b\hat{k}_1)p^2 + (a_2 + b\hat{k}_2)p + b\hat{k}_3}(y(t) - r(t))
\end{aligned} \tag{20}$$

where p is the differential operator. Because a_1, a_2 and b are unknown, the above sensitivity functions cannot be used. Using the MIT rule, we replace a_1, a_2 and b with their estimates $\hat{a}_1$,

$\hat{a}_2$ and $\hat{b}$ in the matching condition (14), i.e. we relate the estimates $\hat{a}_1$, $\hat{a}_2$ and $\hat{b}$ with $\hat{k}_1$, $\hat{k}_2$ and $\hat{k}_3$ using

$$\hat{a}_1 + \hat{b}\hat{k}_1 = a_{m_1},\ \hat{a}_2 + \hat{b}\hat{k}_2 = a_{m_2} \text{ and } \hat{b}\hat{k}_3 = a_{m_3} = b_m \tag{21}$$

and obtain the approximate sensitivity functions as:

$$\begin{aligned} \frac{\partial y}{\partial \hat{k}_1} &\approx \frac{-b}{p^3 + a_{m_1}p^2 + a_{m_2}p + a_{m_3}}\ddot{y}(t) \\ \frac{\partial y}{\partial \hat{k}_2} &\approx \frac{-b}{p^3 + a_{m_1}p^2 + a_{m_2}p + a_{m_3}}\dot{y}(t) \\ \frac{\partial y}{\partial \hat{k}_3} &\approx \frac{-b}{p^3 + a_{m_1}p^2 + a_{m_2}p + a_{m_3}}(y(t) - r(t)) \end{aligned} \tag{22}$$

Finally, the adaptation of $\hat{k}_1$, $\hat{k}_2$ and $\hat{k}_3$ is:

$$\begin{aligned} \frac{d\hat{k}_1}{dt} &= \gamma e \frac{1}{p^3 + a_{m_1}p^2 + a_{m_2}p + a_{m_3}}\ddot{y}(t) \\ \frac{d\hat{k}_2}{dt} &= \gamma e \frac{1}{p^3 + a_{m_1}p^2 + a_{m_2}p + a_{m_3}}\dot{y}(t) \\ \frac{d\hat{k}_3}{dt} &= \gamma e \frac{1}{p^3 + a_{m_1}p^2 + a_{m_2}p + a_{m_3}}(y(t) - r(t)) \end{aligned} \tag{23}$$

where the unknow parameter b in the numerator is absorbed by the gains $\hat{k}_i$.

The equations given by (22) are known as the *sensitivity filters or models*, and can be easily implemented to generate the approximate sensitivity functions for the adaptive law (23). It should be noted that the MRAC based on the MIT rule is locally stable provided the adaptive gain γ is small, the reference input signal r has a small amplitude and sufficient number of frequencies, and the initial conditions $\hat{k}_i(0)$ are close to the nominal values of k_i. For large γ and $\hat{k}_i(0)$ away from the nominal values of k_i, the MIT rule may lead to instability and unbounded signal response.

3.2.2 Conventional MRAC

Consider a multi-input multi-output (MIMO) uncertain linear system (Stepanyan & Krinshnakumar, 2010a):

$$\dot{x}(t) = Ax(t) + B\left[u(t) - f(t)\right],\ x(0) = x_0 \tag{24}$$

where $x \in \Re^n$ and $u \in \Re$ are the state and the control input of the system respectively. $f(t) \in \Re$ is a bounded and piece-wise continuous external disturbance. $A \in \Re^{n\times n}$ and $B \in \Re^{n\times p}$ are unknown constant matrices satisfying the following matching conditions.

Assumption 1. *Given a Hurwitz matrix $A_m \in \Re^{n\times n}$ and a matrix $B_m \in \Re^{n\times p}$ of full column rank, there exists a matrix $K_1 \in \Re^{p\times n}$ and a sign definite matrix $\Lambda \in \Re^{p\times p}$ such that the following equations hold*

$$\begin{aligned} B &= B_m\Lambda \\ A &= A_m - BK_1 \end{aligned} \tag{25}$$

The sign definiteness of Λ corresponds to the conventional sign condition of the high frequency gain matrix of MIMO systems. Without loss of generality, we assume that Λ is positive definite. The rest of the conditions for the existence of an adaptive controller are given by (25).

The control objective is to design a control signal $u(t)$ such that the system tracks the reference model:

$$\dot{x}_m(t) = A_m x_m(t) + B_m N r(t), \;\; x_m(0) = x_0 \tag{26}$$

A_m and B_m are chosen according to performance specifications and satisfy Assumption 1 and $r(t)$ is a bounded and smooth external reference command. The matrix $N = -\left(CA_m^{-1}B_m^{-1}\right)^{-1}$ is chosen such that the output $y_m(t) = Cx_m(t)$ perfectly tracks the reference command $r(t)$. By using (25) and (26), one can note that system (24) can be written in the form:

$$\dot{x}(t) = A_m x(t) + B_m N r(t) + B_m \Lambda \left[u(t) - K_1 x(t) - K_2 r(t) - f(t)\right] \tag{27}$$

where $K_2 = \Lambda^{-1}N$. Hence, choosing the control input $u(t)$ as:

$$u(t) = K_1 x(t) + K_2 r(t) + f(t) \tag{28}$$

translates the system (24) into the reference model (26). The reference model (26) can always be specified from the performance perspective, however the control signal (28) cannot be implemented since the matrices K_1 and K_2 and the vector-function $f(t)$ are assumed to be unknown. Thus, the adaptive control is designed according to the MRAC architecture as:

$$u(t) = \hat{K}_1(t)x(t) + \hat{K}_2(t)r(t) + \hat{f}(t) \tag{29}$$

where $\hat{K}_1(t)$ and $\hat{K}_2(t)$ are the estimates of the ideal control gains K_1 and K_2, and $\hat{f}(t)$ is the estimate of a constant vector $\bar{f}$ that can be referred to as an average value of $f(t)$. These estimates are updated online according to robust adaptive laws:

$$\begin{aligned}
\dot{\hat{K}}_1(t) &= -\gamma B_m^T P\, e(t) x^T(t) + \Psi_1\left(x(t), e(t), \hat{K}_1(t)\right) \\
\dot{\hat{K}}_2(t) &= -\gamma B_m^T P\, e(t) r^T(t) + \Psi_2\left(r(t), e(t), \hat{K}_2(t)\right) \\
\dot{\hat{f}}(t) &= -\gamma B_m^T P\, e(t) + \Psi_3\left(e(t), \hat{f}(t)\right)
\end{aligned} \tag{30}$$

where $\gamma > 0$ is the adaptation rate, $P = P^T > 0$ is the solution of the Lyapunov equation:

$$A_m^T P + P A_m = -Q \tag{31}$$

for some $Q = Q^T > 0$. The terms Ψ_1, Ψ_2 and Ψ_3 represent the robust modifications such as σ-modification, e-modification, projection operator or dead-zone modification (Stepanyan & Krinshnakumar, 2010a). Here $e(t) = x(t) - x_m(t)$ is the tracking error between the system and the reference model.

3.2.3 Modified MRAC

The Modified MRAC (M-MRAC) is proposed in (Stepanyan & Krinshnakumar, 2010a;b). This approach is motivated by the fact that the initial large error in the control gains generates large transient excursions both in system's control and output signals. Therefore, driving the reference model toward the system proportional to the tracking error prevents the system's attempt to aggressively maneuver toward the reference model. In this case, the modified reference model is defined as follows:

$$\dot{x}_m(t) = A_m x_m(t) + B_m N r(t) + \lambda(x(t) - x_m(t)),\ x_m(0) = x_0 \tag{32}$$

where $\lambda > 0$ is a design parameter that specifies the tracking error feedback into the reference model. As the tracking error approaches zero, the reference model (32) approaches its original form, which is called an ideal reference model:

$$\dot{x}^0(t) = A_m x^0(t) + B_m N r(t),\ \ x^0(0) = x_0 \tag{33}$$

The control input u as well as the adaptive laws for $\hat{K}_1$, $\hat{K}_2$ and $\hat{f}$ are the same as in the previous section. It has been shown that the error feedback term $\lambda(x(t) - x_m(t))$ determines the damping in the control signal. Increasing this term makes it possible to increase the learning rate for better transient performance without generating high frequency oscillations in the adaptive system.

3.3 Sliding mode control (SMC)

Closed-loop systems can have some fault-tolerance when the controller gains are carefully chosen to take care of effects of both faults and system uncertainties. Such systems are called passive fault-tolerant control (PFTC) systems. Passive approaches make use of robust control techniques to ensure that a closed-loop system remains insensitive to certain faults using constant controller parameters and without use of on-line fault information. In the case of PFTC, systems continue to operate with the same control gains and structures and the scheme effectiveness depends upon the robustness of the nominal (fault-free) closed-loop system. Systems are made robust to faults by assuming a restrictive repertoire of likely faults and the way in which they affect the control function.

Sliding mode control (SMC) techniques have a strong capability in handling uncertainties and disturbances, which makes them an excellent candidate for passive fault tolerant control system. The design of the SMC for the Qball-X4 starts by expressing the system model (4) in state-space representation (Li, 2011a), (Li et al., 2011b):

$$\begin{bmatrix} \dot{x}_1 \\ \dot{x}_2 \\ \dot{x}_3 \\ \dot{x}_4 \\ \dot{x}_5 \\ \dot{x}_6 \\ \dot{x}_7 \\ \dot{x}_8 \\ \dot{x}_9 \\ \dot{x}_{10} \\ \dot{x}_{11} \\ \dot{x}_{12} \end{bmatrix} = \begin{bmatrix} x_7 \\ x_8 \\ x_9 \\ x_{10} \\ x_{11} \\ x_{12} \\ u_z\left(cosx_5\, sinx_4\, cosx_6 + sinx_5\, sinx_6\right)/m - k_x x_7/m \\ u_z\left(cosx_5\, sinx_4\, sinx_6 - sinx_5\, cosx_6\right)/m - k_y x_8/m \\ u_z\left(cosx_5\, cosx_4\right)/m - g - k_z x_9/m \\ \left(u_\theta + \left(J_y - J_z\right)x_{11}x_{12} - J_T x_{11}\,\Omega - k_\theta x_{10}\right)/J_x \\ \left(u_\phi + \left(J_z - J_x\right)x_{10}x_{12} - J_T x_{10}\,\Omega - k_\phi x_{11}\right)/J_y \\ \left(u_\psi + \left(J_x - J_y\right)x_{10}x_{11} - k_\psi x_{12}\right)/J_z \end{bmatrix} \tag{34}$$

where $[x_1, x_2, x_3, x_4, x_5, x_6, x_7, x_8, x_9, x_{10}, x_{11}, x_{12}]^T = [x, y, z, \theta, \phi, \psi, \dot{x}, \dot{y}, \dot{z}, \dot{\theta}, \dot{\phi}, \dot{\psi}]^T$ and u_z, u_θ, u_ϕ and u_ψ are defined in (9). In order to achieve tracking for the system, the tracking errors need to be defined as following where x_i^d is the desired path for the ith state x_i:

$$e_i = x_i^d - x_i \quad ; \quad i = 1, ..., 6 \tag{35}$$

To guarantee the robustness of the passive fault tolerant control in both fault-free case and fault case, the faults need to be considered during the phase of controller design. Therefore, to design a controller that can handle both fault-free and fault situations, an integral sliding mode technique is employed to further enhance the robustness and to ensure a faster and smoother convergence. The sliding surface is then designed as:

$$s_i = \dot{e}_i + \lambda_i e_i + k_i \int e_i dt \quad ; \quad i = 1, ..., 6 \tag{36}$$

Consider the Lyapunov functions:

$$V_i = \frac{1}{2} s_i^2 \quad ; \quad i = 1, ..., 6 \tag{37}$$

The objective now is to design the control laws so that the time derivatives of the Lyapunov functions are negative. That is:

$$\dot{V}_i = \dot{s}_i s_i < 0 \quad ; \quad i = 1, ..., 6 \tag{38}$$

This latter condition states that the squared distance to the switching surface as measured by s^2 decreases along all system trajectories. However, this condition is not feasible in practice because the switching of real components is not instantaneous and this leads to an undesired phenomenon known as chattering in the direction of the switching surface. Thus, condition (38) is expanded by a boundary layer in which the controller switching is not required as:

$$\dot{s}_i s_i < -\eta |s| \quad ; \quad i = 1, ..., 6 \tag{39}$$

The control inputs should be chosen so that trajectories approach the sliding surface and then stay on it for all future time instants. Thus, the control inputs are expressed as the sum of two terms (Singh & Holé, 2004). The first one, called the equivalent control, is chosen so as to make $\dot{s}_i = 0$ when $s_i = 0$. Taking the time derivative of (36), the dynamics of the sliding surfaces can be obtained as following:

$$\begin{aligned}
\dot{s}_1 &= \ddot{x}_1^d - u_z u_x + k_x x_7/m + \lambda_1 \dot{e}_1 + k_1 e_1 \\
\dot{s}_2 &= \ddot{x}_2^d - u_z u_y + k_y x_8/m + \lambda_2 \dot{e}_2 + k_2 e_2 \\
\dot{s}_3 &= \ddot{x}_3^d - u_z (\cos x_5 \cos x_4)/m + g + k_z x_9/m + \lambda_3 \dot{e}_3 + k_3 e_3 \\
\dot{s}_4 &= \ddot{x}_4^d - (u_\theta + (J_y - J_z) x_{11} x_{12} - J_T x_{11} \Omega - k_\theta x_{10})/J_x + \lambda_4 \dot{e}_4 + k_4 e_4 \\
\dot{s}_5 &= \ddot{x}_5^d - (u_\phi + (J_z - J_x) x_{10} x_{12} - J_T x_{10} \Omega - k_\phi x_{11})/J_y + \lambda_5 \dot{e}_5 + k_5 e_5 \\
\dot{s}_6 &= \ddot{x}_6^d - (u_\psi + (J_x - J_y) x_{10} x_{11} - k_\psi x_{12})/J_z + \lambda_6 \dot{e}_6 + k_6 e_6
\end{aligned} \tag{40}$$

where $u_x = (\cos x_5 \sin x_4 \cos x_6 + \sin x_5 \sin x_6)/m$ and $u_y = (\cos x_5 \sin x_4 \sin x_6 - \sin x_5 \cos x_6)/m$ are defined as virtual inputs. By setting $\dot{s}_i = 0$ in

(40) the equivalent control inputs can be derived as:

$$\begin{aligned}
u_x^{eq} &= \frac{1}{u_z}\left(\ddot{x}_1^d + k_x x_7/m + \lambda_1 \dot{e}_1 + k_1 e_1\right)\\
u_y^{eq} &= \frac{1}{u_z}\left(\ddot{x}_2^d + k_y x_8/m + \lambda_2 \dot{e}_2 + k_2 e_2\right)\\
u_z^{eq} &= m\left(\ddot{x}_3^d + g + k_z x_9/m + \lambda_3 \dot{e}_3 + k_3 e_3\right)/\left(cosx_5\ cosx_4\right)\\
u_\theta^{eq} &= J_x\left(\ddot{x}_4^d + \lambda_4 \dot{e}_4 + k_4 e_4\right) - \left(J_y - J_z\right)x_{11}x_{12} + J_T x_{11}\,\Omega + k_\theta x_{10}\\
u_\phi^{eq} &= J_y\left(\ddot{x}_5^d + \lambda_5 \dot{e}_5 + k_5 e_5\right) - \left(J_z - J_x\right)x_{10}x_{12} + J_T x_{10}\,\Omega + k_\phi x_{11}\\
u_\psi^{eq} &= J_z\left(\ddot{x}_6^d + \lambda_6 \dot{e}_6 + k_6 e_6\right) - \left(J_x - J_y\right)x_{10}x_{11} + k_\psi x_{12}
\end{aligned} \tag{41}$$

The second term of the control inputs is chosen to tackle uncertainties in the system and to introduce reaching law. One can achieve a proportional $(l_i s_i)$ plus constant $(m_i sign(s_i))$ rate reaching law by selecting the second term as:

$$\begin{aligned}
u_x^* &= \frac{1}{u_z}\left(l_1 s_1 + m_1 sign(s_1)\right)\\
u_y^* &= \frac{1}{u_z}\left(l_2 s_2 + m_2 sign(s_2)\right)\\
u_z^* &= \frac{m}{\left(cosx_5\ cosx_4\right)}\left(l_3 s_3 + m_3 sign(s_3)\right)\\
u_\theta^* &= J_x\left(l_4 s_4 + m_4 sign(s_4)\right)\\
u_\phi^* &= J_y\left(l_5 s_5 + m_5 sign(s_5)\right)\\
u_\psi^* &= J_z\left(l_6 s_6 + m_6 sign(s_6)\right)
\end{aligned} \tag{42}$$

where l_i and m_i $(i = 1, ..., 6)$ are positive control gains. The total control inputs $u = u^{eq} + u^*$ are then (Li, 2011a), (Li et al., 2011b):

$$\begin{aligned}
u_x &= \frac{1}{u_z}\left(\ddot{x}_1^d + k_x x_7/m + \lambda_1 \dot{e}_1 + k_1 e_1 + l_1 s_1 + m_1 sign(s_1)\right)\\
u_y &= \frac{1}{u_z}\left(\ddot{x}_2^d + k_y x_8/m + \lambda_2 \dot{e}_2 + k_2 e_2 + l_2 s_2 + m_2 sign(s_2)\right)\\
u_z &= \frac{m}{\left(cosx_5\ cosx_4\right)}\left(\ddot{x}_3^d + g + k_z x_9/m + \lambda_3 \dot{e}_3 + k_3 e_3 + l_3 s_3 + m_3 sign(s_3)\right)\\
u_\theta &= J_x\left(\ddot{x}_4^d + \lambda_4 \dot{e}_4 + k_4 e_4 + l_4 s_4 + m_4 sign(s_4)\right) - \left(J_y - J_z\right)x_{11}x_{12} + J_T x_{11}\,\Omega + k_\theta x_{10}\\
u_\phi &= J_y\left(\ddot{x}_5^d + \lambda_5 \dot{e}_5 + k_5 e_5 + l_5 s_5 + m_5 sign(s_5)\right) - \left(J_z - J_x\right)x_{10}x_{12} + J_T x_{10}\,\Omega + k_\phi x_{11}\\
u_\psi &= J_z\left(\ddot{x}_6^d + \lambda_6 \dot{e}_6 + k_6 e_6 + l_6 s_6 + m_6 sign(s_6)\right) - \left(J_x - J_y\right)x_{10}x_{11} + k_\psi x_{12}
\end{aligned} \tag{43}$$

It is straightforward to verify the negativity of condition (39). For illustration purpose, taking for example u_x of (43) and plugging it into $\dot{s}_1$ of (40) gives:

$$\dot{s}_1 = -l_1 s_1 - m_1 sign(s_1) \tag{44}$$

Multiplying both sides of the previous equation by s_1:

$$s_1\dot{s}_1 = -l_1 s_1^2 - m_1 |s_1| \tag{45}$$

and properly choosing control gains l_1 and m_1 allows to verify (39). The same procedure can be followed for the other control inputs.

The elimination of the chattering effect produced by the discontinuous function *sign* can be attained by using a saturation function *sat* instead of the *sign* function. This saturation function is defined as follows:

$$sat(s_i) = \begin{cases} \delta_b & \text{if} \quad s_i \geq \delta_b \\ s_i & \text{if} \quad -\delta_b < s_i < \delta_b \\ -\delta_b & \text{if} \quad s_i \leq -\delta_b \end{cases} \tag{46}$$

where δ_b is the boundary of the saturation function and is set small enough.

3.4 Backstepping control (BSC)

Backstepping design refers to "step back" to the control input, and a major advantage of backstepping design is its flexibility to avoid cancellation of useful nonlinearities and pursue the objectives of stabilization and tracking, rather than those of linearization. Recursively constructed backstepping controller (BSC) employs the control Lyapunov function to guarantee the global stability (Krstic et al., 1995).

In order to illustrate the BSC techniques, consider the following nonlinear system:

$$\begin{aligned} \dot{x} &= f(x) + g(x)\xi \\ \dot{\xi} &= u \end{aligned} \tag{47}$$

where x is the state vector and ξ is the control input. As a first step, define the tracking error between the actual value x and its desired value x^d as follows:

$$e = x^d - x \tag{48}$$

As a second step, define the following states:

$$\begin{aligned} x_1 &= x \\ x_2 &= \dot{x}_1 = \dot{x} \\ z_1 &= e = x^d - x = x^d - x_1 \end{aligned} \tag{49}$$

The key idea of BSC is to choose certain variables as virtual controls. Assuming x_2 is the virtual control variable and $a(z_1)$ is the function which makes $z_1 \to 0$, define $\overline{z}_1$ as:

$$\overline{z}_1 = a(z_1) - x_2 \tag{50}$$

The time derivative of z_1 is:

$$\dot{z}_1 = \dot{x}^d - \dot{x}_1 = \dot{x}^d - x_2 = \dot{x}^d + \overline{z}_1 - a(z_1) \tag{51}$$

Let us choose $a(z_1)$ as $a(z_1) = \dot{x}^d + k_1 z_1$ with $k_1 > 0$ and define the Lyapunov function:

$$V_1 = \frac{1}{2} z_1^2 \tag{52}$$

One can see that the time derivative of V_1 is:

$$\dot{V}_1 = z_1\dot{z}_1 = z_1\left(\dot{x}^d + \bar{z}_1 - a(z_1)\right) = z_1\left(\bar{z}_1 - k_1 z_1\right) = -k_1 z_1^2 + z_1\bar{z}_1 \tag{53}$$

Clearly if $\bar{z}_1 = 0$, then $\dot{V}_1 = -k_1 z_1^2$ and z_1 is guaranteed to converge to zero asymptotically. Then, the next step is to define a Lyapunov function so as to make $\bar{z}_1 \to 0$:

$$V_2 = V_1 + \frac{1}{2}\bar{z}_1^2 \tag{54}$$

The time derivative of V_2 is:

$$\dot{V}_2 = \dot{V}_1 + \bar{z}_1\dot{\bar{z}}_1 = -k_1 z_1^2 + z_1\bar{z}_1 + \bar{z}_1\left(\ddot{x}^d + k_1\dot{z}_1 - \dot{x}_2\right) \tag{55}$$

To obtain $\dot{V}_2 < 0$, $\dot{x}_2$ is chosen as:

$$\dot{x}_2 = \ddot{x}^d + \left(1 - k_1^2\right) z_1 + \left(k_1 + k_2\right)\bar{z}_1 \tag{56}$$

with $k_1 > 0, k_2 > 0, z_1 = x^d - x_1$ and $\bar{z}_1 = \dot{x}^d + k_1 z_1 - x_2$.

The application of the method above to the state-space representation (34) of the Qball-X4 model gives the control inputs (Zhang, 2010a), (Zhang et al., 2010b):

$$\begin{aligned}
u_x &= \frac{1}{u_z}\left(\ddot{x}_1^d + k_x x_7/m + \left(1-k_1^2\right) z_1 + \left(k_1+k_2\right)\bar{z}_1\right)\\
u_y &= \frac{1}{u_z}\left(\ddot{x}_2^d + k_y x_8/m + \left(1-k_1^2\right) z_2 + \left(k_1+k_2\right)\bar{z}_2\right)\\
u_z &= \frac{m}{(\cos x_5 \, \cos x_4)}\left(\ddot{x}_3^d + g + k_z x_9/m + \left(1-k_1^2\right) z_3 + \left(k_1+k_2\right)\bar{z}_3\right)\\
u_\theta &= J_x\left(\ddot{x}_4^d + \left(1-k_1^2\right) z_4 + \left(k_1+k_2\right)\bar{z}_4\right) - \left(J_y - J_z\right) x_{11}x_{12} + J_T x_{11}\,\Omega + k_\theta x_{10}\\
u_\phi &= J_y\left(\ddot{x}_5^d + \left(1-k_1^2\right) z_5 + \left(k_1+k_2\right)\bar{z}_5\right) - \left(J_z - J_x\right) x_{10}x_{12} + J_T x_{10}\,\Omega + k_\phi x_{11}\\
u_\psi &= J_z\left(\ddot{x}_6^d + \left(1-k_1^2\right) z_6 + \left(k_1+k_2\right)\bar{z}_6\right) - \left(J_x - J_y\right) x_{10}x_{11} + k_\psi x_{12}
\end{aligned} \tag{57}$$

with $u_x = \left(\cos x_5 \, \sin x_4 \, \cos x_6 + \sin x_5 \, \sin x_6\right)/m$ and $u_y = \left(\cos x_5 \, \sin x_4 \, \sin x_6 - \sin x_5 \, \cos x_6\right)/m$ defined as virtual inputs and $z_i = x_i^d - x_i$ and $\bar{z}_i = \dot{x}_i^d + k_1 z_i - \dot{x}_i$ $(i = 1, ..., 6)$.

3.5 Model predictive control (MPC)

Model Predictive Control (MPC) is a promising tool for fault tolerant control applications (Maciejowski & Jones, 2003) due to its prominent capabilities such as constraint handling, flexibility to changes in the process dynamics and applicability to nonlinear dynamics. Since MPC recalculates the control signal at each sampling time, any change in the process dynamics can be reflected simply into the control signal calculation. Also, the constraint handling capability of MPC allows system working close to the boundaries of the tight post-failure operation envelope. The drawback of MPC is that similarly to most of the control techniques, it needs an almost explicit model of the system to calculate a stabilizing control signal. On the other hand, the abrupt changes in the model parameters, due to failure, cannot be predicted beforehand and an online data-driven parameter estimation methodology is required to extract the post-failure model parameters from online input/output data. In other words, an FDD module is required to provide information about the occurring faults to allow MPC to consider faults.

3.5.1 Notation and terminology

In the MPC, also known as Receding Horizon Control (RHC), a cost function is minimized over a future prediction horizon time step denoted by N, subject to the dynamical constraints. The first control input in the sequence is applied to the plant until the next update is available. The discrete timing is shown by k where $k \in \mathbb{N}$. The state vectors are introduced as follows:

- $x(k)$: actual state vector at time step k.
- $x_k(j)$: predicted state vector at time step $k+j$ computed at time step k ($j \in \{0,1,...,N\}$).

Similar notation is used for the control input vector u. Also the sequence of the state/input vector over the prediction horizon is called the state/input trajectory and is shown as follows:

$$\begin{aligned} x_k(.) &= \{x_k(j) | j = 0,1,2,...,N\} \\ u_k(.) &= \{u_k(j) | j = 0,1,2,...,N-1\} \end{aligned} \tag{58}$$

3.5.2 Fault-free MPC formulation

The cost function at time step k is defined as follows:

$$J(x_k(.), u_k(.)) = \sum_{j=0}^{N-1} \left(\|x_k(j) - x^d\|_Q^2 + \|u_k(j)\|_R^2 \right) + \|x_k(N) - x^d\|_P^2 \tag{59}$$

where $\|x\|_Q^2 = x^T Q x$, and $P > 0$, $Q > 0$ and $R > 0$ are symmetric matrices. x^d is the desired or reference state of x. Then the MPC problem at each step time k is defined as follows:

Problem 1. *Calculate:*

$$J^*(x(k)) = \min_{\{u_k(.), x_k(.)\}} J(x_k(.), u_k(.)) \tag{60}$$

Subject to:

$$x_k(j+1) = f(x_k(j), u_k(j)) \quad ; \quad x_k(0) = x(k) \tag{61}$$

$$x_k(j) \in \mathbb{X} \tag{62}$$

$$u_k(j) \in \mathbb{U} \tag{63}$$

$$x_k(N) \in \mathbb{X}_t \tag{64}$$

for $j = 0,1,...,N-1$.

$\mathbb{X} \subseteq \mathbb{R}^n$, $\mathbb{U} \subseteq \mathbb{R}^m$ and $\mathbb{X}_t \subseteq \mathbb{X}$ denote the set of admissible states, inputs and terminal states (terminal region) respectively. J^* denotes the optimal value of the cost function J. This MPC formulation is based on the quasi-infinite model predictive control (Chen & Allgower, 1998). At each time step k, MPC generates the input and state trajectories, by solving Problem 1. After generating these trajectories, MPC controller applies only the first computed control input, i.e. $u_k(0)$ to the system. The following algorithm presents the online implementation of Problem 1 (Izadi, 2009a), (Izadi et al., 2011b).

Algorithm 1. *Given* $x(0)$ *and* x^d, *do:*

1. $k = 0$.
2. *Measure* $x(k)$.

3. *Solve Problem 1 and generate* $u_k(.)$ *and* $x_k(.)$.
4. *Apply* $u_k(0)$ *to the system.*
5. *Set* $k = k + 1$ *and GOTO Step #2.*

This algorithm is repeated for $k = 0, 1, ..., \infty$. In step #2, if full state measurement is not available, then a state estimator (observer) can be used to provide the state estimate.

3.5.3 Fault-tolerant MPC

In the presence of an actuator fault, the system dynamics can be rewritten as follows:

$$x(k+1) = f(x(k), \alpha(k)u(k)) \tag{65}$$

where α captures the fault information and is called the fault parameter matrix which determines the fault severity. α is a diagonal matrix, i.e. $\alpha = diag(\alpha_1, \alpha_2, ..., \alpha_m)$ where m is the number of inputs ($m = 4$ for the quadrotor helicopter) and the scalar α_i denotes the amplitude of the fault in the ith actuator. $\alpha_i = 1$ denotes a healthy actuator, $\alpha_i = 0$ denotes a complete loss of actuator effectiveness and $0 < \alpha_i < 1$ represents a partial loss. Taking into consideration the model representation (65) after fault occurrence, the fault-tolerant MPC problem can be defined as follows (Izadi et al., 2011b):

Problem 2. *Calculate:*

$$J^*(x(k)) = \min_{\{u_k(.), x_k(.)\}} J(x_k(.), u_k(.)) \tag{66}$$

Subject to:

$$x_k(j+1) = f(x_k(j), \alpha u_k(j)) \quad ; \quad x_k(0) = x(k) \tag{67}$$

$$x_k(j) \in \mathbb{X} \tag{68}$$

$$u_k(j) \in \mathbb{U} \tag{69}$$

$$x_k(N) \in \mathbb{X}_t \tag{70}$$

for $j = 0, 1, ..., N - 1$.

One can see that the difference between Problem 1 and Problem 2 resides in the consideration of the fault matrix α in (67). Clearly, solving Problem 2 requires an FDD module for fault identification and thus the algorithm for solving Problem 2 is similar to Algorithm1 except that an FDD module is needed to run parallelly to provide an estimation of α.

3.6 Flatness-based trajectory planning/re-planning (FTPR)

This section does not present an FTC method but a trajectory planning/re-planning approach that can be combined with any FTC module to provide a better management of the system resources after fault occurrence. This is achieved by redefining the reference trajectories to be followed by the damaged system depending on the fault severity. This approach is equivalent to the reconfiguration of command governor block in Figure 3.

3.6.1 Flatness notion

The flatness can be defined as follows. A dynamic system:

$$\begin{aligned} \dot{x} &= f(x,u) \\ y &= h(x) \end{aligned} \tag{71}$$

with $x \in \Re^n$ and $u \in \Re^m$, is flat if and only if there exist variables $F \in \Re^m$ called the flat outputs such that: $x = \Xi_1(F, \dot{F}, ..., F^{(n-1)})$, $y = \Xi_2(F, \dot{F}, ..., F^{(n-1)})$ and $u = \Xi_3(F, \dot{F}, ..., F^{(n)})$. Ξ_1, Ξ_2 and Ξ_3 are three smooth mappings and $F^{(i)}$ is the ith derivative of F. The parameterization of the control inputs u in function of the flat outputs F plays a key role in the trajectory planning problem: the nominal control inputs to be applied during a mission can be expressed in function of the desired trajectories. Thus allowing to tune the profile of the trajectories for keeping the applied control inputs below the actuator limits (Chamseddine et al., 2011b).

For the quadrotor UAV simplified model given in (6), one can see that the outputs to be controlled can be chosen as flat outputs. Thus, system (6) is flat with flat outputs $F_1 = z$, $F_2 = x$, $F_3 = y$ and $F_4 = \psi$. In addition to x, y, z and ψ, the parameterization of θ and ϕ in function of the flat outputs is:

$$\begin{aligned} \theta &= \frac{\ddot{F}_2}{g}; \\ \phi &= -\frac{\ddot{F}_3}{g} \end{aligned} \tag{72}$$

The parameterization of the control inputs in function of the flat outputs is:

$$\begin{aligned} u_z &= m\left(\ddot{F}_1 + g\right); \\ u_\theta &= J_x \frac{F_2^{(4)}}{g}; \\ u_\phi &= -J_y \frac{F_3^{(4)}}{g}; \\ u_\psi &= J_z \ddot{F}_4 \end{aligned} \tag{73}$$

3.6.2 Reference trajectory design

Let's define F_i^* as the reference trajectories for the flat output F_i. Several methods can be used to design F_i^*. In this work we employ the Bézier polynomial function. A general Bézier polynomial function of degree n is:

$$F = a_n t^n + a_{n-1} t^{n-1} + ... + a_2 t^2 + a_1 t + a_0 \tag{74}$$

where t is the time and a_i $(i = 0, ..., n)$ are constant coefficients to be calculated in function of the initial and final conditions. It is clear that the larger is n, the smoother is the reference trajectory. However, calculations for trajectory planning become heavier as n increases. For the quad-rotor UAV, it can be seen in (73) that u_z is function of $\ddot{F}_1$, u_θ of $F_2^{(4)}$, u_ϕ of $F_3^{(4)}$ and

u_ψ of $\ddot{F}_4$. The relative degrees are then $r_1 = 2$, $r_2 = 4$, $r_3 = 4$ and $r_4 = 2$. Smooth control inputs can be obtained if one can impose $F_i^{(0)}$ up to $F_i^{(n_i)}$ at the initial and final time (i.e. up to $F_i^{(n_i)}(t_0)$ and $F_i^{(n_i)}(t_f)$) where $n_i = 2$ for $i = 1,4$ and $n_i = 4$ for $i = 2,3$. To this end, we employ a Bézier polynomial function of degree 5 for F_1 and F_4 and a Bézier polynomial function of degree 9 for F_2 and F_3. The reference trajectories are then:

$$F_i^* = a_5^i t^5 + a_4^i t^4 + a_3^i t^3 + a_2^i t^2 + a_1^i t + a_0^i \ ; \ (i = 1,4) \tag{75}$$

and

$$F_i^* = a_9^i t^9 + a_8^i t^8 + \cdots + a_2^i t^2 + a_1^i t + a_0^i \ ; \ (i = 2,3) \tag{76}$$

The coefficients a_j^i ($i = 1,4$ and $j = 0,...,5$) are calculated to verify the initial conditions $F_i(t_0)$, $\dot{F}_i(t_0)$, $\ddot{F}_i(t_0)$ and the final conditions $F_i(t_f)$, $\dot{F}_i(t_f)$, $\ddot{F}_i(t_f)$. The coefficients a_j^i ($i = 2,3$ and $j = 0,...,9$) are calculated to verify the initial conditions $F_i(t_0)$, $\dot{F}_i(t_0)$, $\ddot{F}_i(t_0)$, $F_i^{(3)}(t_0)$, $F_i^{(4)}(t_0)$ and the final conditions $F_i(t_f)$, $\dot{F}_i(t_f)$, $\ddot{F}_i(t_f)$, $F_i^{(3)}(t_f)$, $F_i^{(4)}(t_f)$. t_0 and t_f are respectively the initial and the final instants of the mission.

3.6.3 Trajectory planning

The trajectory planning consists in driving the quadrotor from an initial to a final position. The initial and final conditions as well as the initial time t_0 are all known and the only unknown is the final time of the mission t_f. Thus, the trajectory planning tends to tune the profile of the trajectory (by tuning t_f) so that to drive the system along the desired trajectory without violating actuator constraints. Thanks to flatness, it is possible to find the relation between the applied control inputs and the reference trajectories. According to (73) the nominal control inputs to be applied along the reference trajectories are:

$$\begin{aligned} u_z^* &= m\left(\ddot{F}_1^* + g\right); \\ u_\theta^* &= J_x \frac{F_2^{(4)*}}{g}; \\ u_\phi^* &= -J_y \frac{F_3^{(4)*}}{g}; \\ u_\psi^* &= J_z \ddot{F}_4^* \end{aligned} \tag{77}$$

Since the quadrotor is assumed to have a zero yaw angle ($\psi = 0$), then $\ddot{F}_4^* = 0 \ \forall\, t \in [t_0, t_f]$. It is also assumed that the quadrotor is not changing altitude and thus $\ddot{F}_1^* = 0 \ \forall\, t \in [t_0, t_f]$. In this case, the nominal control inputs are:

$$\begin{aligned} u_z^* &= mg; \\ u_\theta^* &= J_x \frac{F_2^{(4)*}}{g}; \\ u_\phi^* &= -J_y \frac{F_3^{(4)*}}{g}; \\ u_\psi^* &= 0 \end{aligned} \tag{78}$$

The system is required to move from an initial position $F_i(t_0) = 0$ with initial conditions $\dot{F}_i(t_0) = \ddot{F}_i(t_0) = F_i^{(3)}(t_0) = F_i^{(4)}(t_0) = 0$ to a final position $F_i(t_f) \neq 0$ with final conditions $\dot{F}_i(t_f) = \ddot{F}_i(t_f) = F_i^{(3)}(t_f) = F_i^{(4)}(t_f) = 0$ $(i = 2,3)$. Without loss of generality, by setting $t_0 = 0$ one can write (76) as

$$F_i^* = 70F_i(t_f)\frac{t^9}{t_f^9} - 315F_i(t_f)\frac{t^8}{t_f^8} + 540F_i(t_f)\frac{t^7}{t_f^7} - 420F_i(t_f)\frac{t^6}{t_f^6} + 126F_i(t_f)\frac{t^5}{t_f^5} \; ; \; (i = 2,3) \quad (79)$$

Thus, it is possible to calculate $u_\theta^* = J_x F_2^{(4)*}/g$ and $u_\phi^* = -J_y F_3^{(4)*}/g$ since

$$F_i^{(4)*} = c_1 F_i(t_f)\frac{t^5}{t_f^9} - c_2 F_i(t_f)\frac{t^4}{t_f^8} + c_3 F_i(t_f)\frac{t^3}{t_f^7} - c_4 F_i(t_f)\frac{t^2}{t_f^6} + c_5 F_i(t_f)\frac{t}{t_f^5} \; ; \; (i = 2,3) \quad (80)$$

where c_j $(j = 1,...,5)$ are constants that can be easily derived from (79) and are not given here for simplicity. From (9), (78) and (80), it is possible to determine the nominal PWM inputs to be applied along the references trajectories, which are:

$$u_{1/2}^* = \frac{mg}{4K} \pm \frac{F_2(t_f)J_x\Delta}{2gKLt_f^9} \text{ and } u_{4/3}^* = \frac{mg}{4K} \pm \frac{F_3(t_f)J_y\Delta}{2gKLt_f^9} \quad (81)$$

where $\Delta = \left(c_1 t^5 - c_2 t_f t^4 + c_3 t_f^2 t^3 - c_4 t_f^3 t^2 + c_5 t_f^4 t\right)$. Since $F_2(t_f)$ and $F_3(t_f)$ are known, one must find the minimal t_f so that $u_i^* < u_{max}$ where u_{max} is the maximal PWM that corresponds to the maximal thrust that can be generated by a rotor. One way to determine t_f is to analytically calculate when the extrema of u_i^* take place and what are their values (where the extrema collectively denote the maxima and the minima of a function). For this purpose, it is necessary to calculate the derivative of u_i^* in (81) with respect to time and then solve for t the following equation:

$$\frac{du_i^*}{dt} = 0 \; ; \; (i = 1,...,4) \quad (82)$$

For a fixed i $(i = 1,...,4)$, each equation du_i^*/dt is a polynomial function of fourth degree and thus has four extrema which take place at:

$$\frac{t_f}{2}\left(1 \pm \sqrt{\frac{3}{7} - \frac{2}{35}\sqrt{30}}\right) \text{ and } \frac{t_f}{2}\left(1 \pm \sqrt{\frac{3}{7} + \frac{2}{35}\sqrt{30}}\right) \quad (83)$$

The values of the four extrema can be obtained by replacing t of (81) by the values obtained in (83). As an example, the values of the four extrema of u_1^* of the first rotor are:

$$u_{1_{Ext}}^* = \left(\frac{mg}{4K} \pm c_6\frac{F_2(t_f)J_x}{gKLt_f^4}; \frac{mg}{4K} \pm c_7\frac{F_2(t_f)J_x}{gKLt_f^4}\right) \quad (84)$$

with c_6 and c_7 two constants. The determination of t_f consists finally in solving (84) so that the extrema of the nominal PWM input verifies:

$$u_{1_{Ext}}^* \leq \rho u_{max} \quad (85)$$

Since the above study is based on the simplified model (6), the constant $0 \leq \rho \leq 1$ is introduced to create some safety margin and to robustify the obtained solution with respect to model uncertainties. Four solutions can then be obtained:

$$t_f \geq \left(\left(\pm c_6 \frac{F_2(t_f) J_x}{gKL\left(\rho u_{max} - \frac{mg}{4K}\right)} \right)^{1/4} ; \left(\pm c_7 \frac{F_2(t_f) J_x}{gKL\left(\rho u_{max} - \frac{mg}{4K}\right)} \right)^{1/4} \right) \tag{86}$$

The four extrema of u_2^*, u_3^* and u_4^* have the same structure as for u_1^* with different values for the constants c_6 and c_7. In conclusion, the quadrotor UAV will have 16 extrema (four per motor) and therefore 16 solutions are obtained and the maximal one among them is to be considered.

3.6.4 Trajectory re-planning

As for the fault-free case, the trajectory re-planning consists in determining the minimal time of the mission t_f so that the actuator constraints are not violated. For the damaged UAV, this is written as:

$$u_i^* \leq \rho(1 - \delta_i) u_{max} \ ; \ (i = 1, ..., 4) \tag{87}$$

where δ_i represents the loss of effectiveness in the ith rotor. $\delta_i = 0$ denotes a healthy rotor, $\delta_i = 1$ denotes a complete loss of the rotor and $0 < \delta_i < 1$ represents a partial loss of control effectiveness. Clearly, trajectory re-planning in the fault case requires the knowledge of the fault amplitude δ_i. Thus, a fault detection and diagnosis module is needed to detect, isolate and identify the fault. Unlike trajectory planning, the initial conditions are not zero at the re-planning instant i.e. $\dot{F}_i(t_{rep}) \neq 0, \ddot{F}_i(t_{rep}) \neq 0, F_i^{(3)}(t_{rep}) \neq 0$ and $F_i^{(4)}(t_{rep}) \neq 0$ where t_{rep} is the re-planning time. Therefore, it is difficult to analytically calculate t_f as in the previous section since the expressions are much more complicated and thus another method is needed to solve the problem. Starting from the idea that the larger t_f is, the smaller are the PWM inputs u_i, we propose the following algorithm to calculate t_f:

1. Develop the expressions of $u_i^*(t)$ (similar to the procedure in the previous section) with nonzero initial conditions.
2. Start with an initial guess of the mission time t_f.
3. For $t = 0 : T_e : t_f$ calculate $u_i^*(t)$ and determine the maximum value $u_{i_{max}}^*$ among all the $u_i^*(t)$. T_e is a sampling period.
4. Determine the error E_i between $u_{i_{max}}^*$ and the maximal allowable PWM input (ρu_{max} for the healthy UAVs and $\rho(1 - \delta_i) u_{max}$ for the damaged one).
5. If the error is smaller than a predefined threshold, exit the algorithm and the solution is the current t_f. If not, update the current t_f as follows $t_f \leftarrow t_f + t_f E / \nu$ with ν is a positive constant.
6. Repeat from Step #3.

As can be seen in Step #5, if the error E_i is positive then $u_{i_{max}}^*$ is larger than the maximal allowable PWM input and thus it is necessary to increase t_f. If the error is negative then $u_{i_{max}}^*$ is smaller than the maximal allowable PWM input and thus it is necessary to decrease t_f. The constant ν must be carefully selected since it affects the convergence speed of the algorithm.

4. Simulation and experimental testing results

The approaches that have been experimentally tested on the Qball-X4 are built using Matlab/Simulink and downloaded on the Gumstix emdedded computer to be run on-board with a frequency of 200 Hz. The experiments are taking place indoor in the absence of GPS signals and thus the OptiTrack camera system from NaturalPoint is employed to provide the system position in the 3D space. Some photos of the NAV Lab of Concordia University are shown in Figure 5 and many videos related to the above approaches as well as other videos can be watched online on `http://www.youtube.com/user/NAVConcordia`.

Fig. 5. The NAV Lab of Concordia University.

4.1 GS-PID experimental results

Starting with the GS-PID, an 18% loss of overall power of all motors is considered where the Qball-X4 is requested to track a one meter square trajectory tracking. The fault-free case is shown in the left-hand side plot of Figure 6 whereas the middle one shows a deviation from the desired trajectory after the fault occurrence when the switching between the PID gains is taking place after 0.5 s of fault occurrence. Better tracking performance can be achieved with shorter switching time which requires fast and correct fault detection. The right-hand plot of Figure 6 demonstrates better performance when the switching is done without time delay.

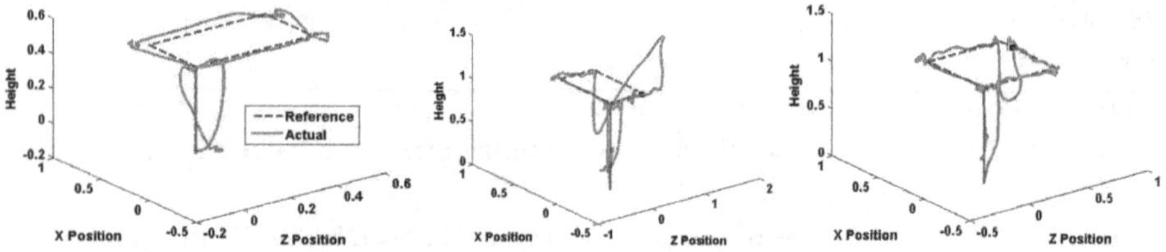

Fig. 6. Fault-free case and 18% loss in the total thrust with and without time delay.

4.2 MRAC experimental results

As for the MRAC, a partial effectiveness loss of 30% and 40% are simulated in the total thrust u_z. The faults are injected at $t = 40$ s and the experimental results are illustrated in Figure 7. This kind of fault induces a loss in the height z without significant effect on x, y or ψ directions. One can see that all the three MRAC techniques give better performance than the baseline LQR controller and that the conventional MRAC (C-MRAC) is the best among the MRAC techniques.

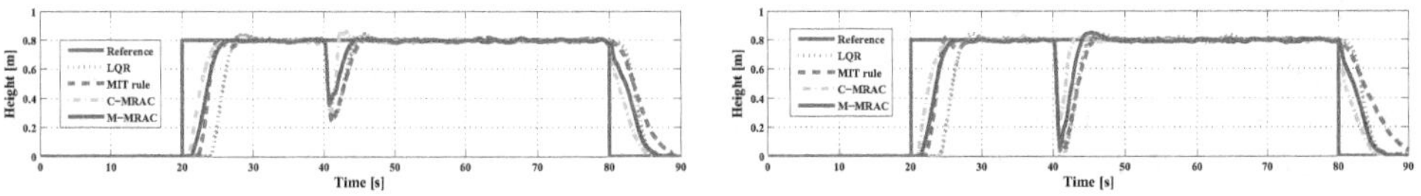

Fig. 7. System behavior with a simulated fault of 30% and 40% loss in the total thrust.

The MRAC has been also tested in the presence of a partial damage of the 4th propeller. This type of faults is more realistic than the simulated faults given above. To accommodate a partial damage of the propeller, the controller tries to speed up the 4th rotor so that to produce the same lift as for the fault-free case. Thus, the maximal speed that a rotor can reach is a critical factor that does not play a role in simulated faults. Figure 8 shows the system behavior along the z and y directions where the first row corresponds to 12% damage and the second to 16%. One can see that the performance obtained with the C-MRAC is better than that of the LQR.

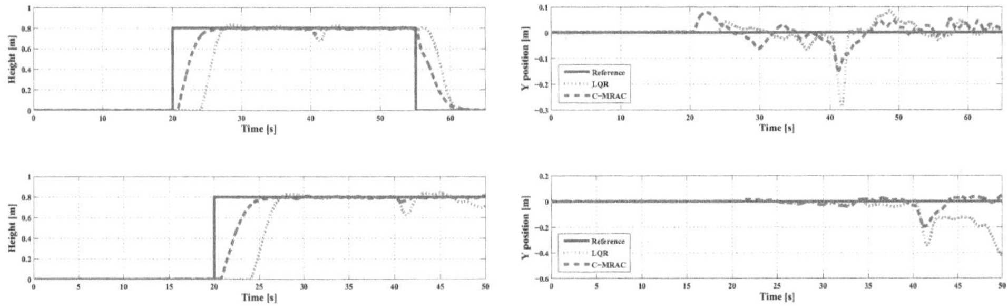

Fig. 8. System behavior with a damage of 12% and 16% of the fourth propeller.

4.3 SMC experimental results

The objective in the SMC is to track a square trajectory of 1.5 $m \times 1.5$ m. Figure 9 shows the system evolution along the x, y and z directions when the SMC-based PFTC is experimentally applied to the Qball-X4. The SMC-based PFTC is giving good performance where very small deviation in position z can be observed at 20 s due to a partial damage in the 4th propeller.

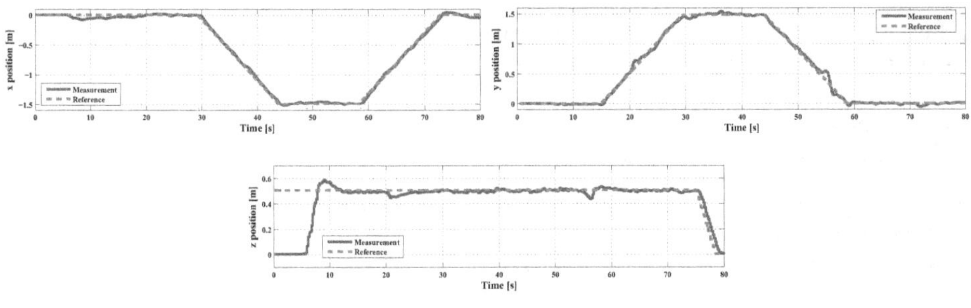

Fig. 9. Qball-X4 evolution along the x, y and z directions using SMC-based PFTC.

4.4 BSC simulation results

In the BSC approach, the system is required to follow a circular trajectory and the the actuator faults are injected at time $t = 5$ s. Simulations are carried out with different control gains $k_1 = 1$, $k_2 = 3$ and $k_1 = 5$, $k_2 = 30$. Figures 10(a) and 10(b) show a comparison between the fault-free case and the fault-case of 50% loss of control effectiveness for both position and angle tracking errors. The control gains used in Figures 10(a) and 10(b) are $k_1 = 1$ and $k_2 = 3$. One can see that in the fault case, the position tracking errors in the x and y directions change slightly whereas the z tracking error is greatly affected. The roll, pitch and yaw tracking errors are also affected as can be seen in Figure 10(b). With higher control gains $k_1 = 5$ and $k_2 = 30$, it is possible to reduce fault effects on tracking errors as can be seen in Figures 10(c) and 10(d).

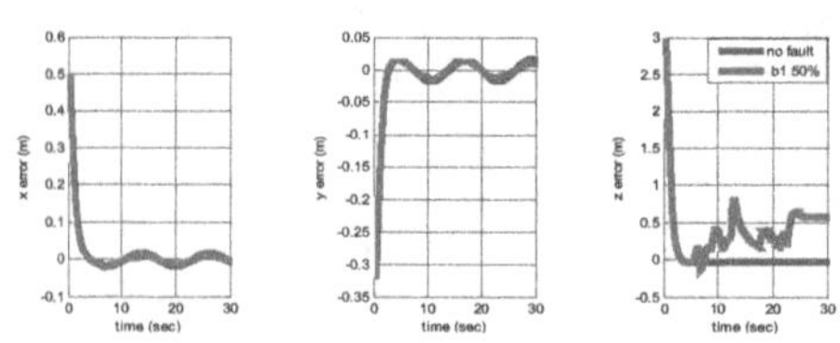

(a) Position errors with $k_1 = 1/k_2 = 3$

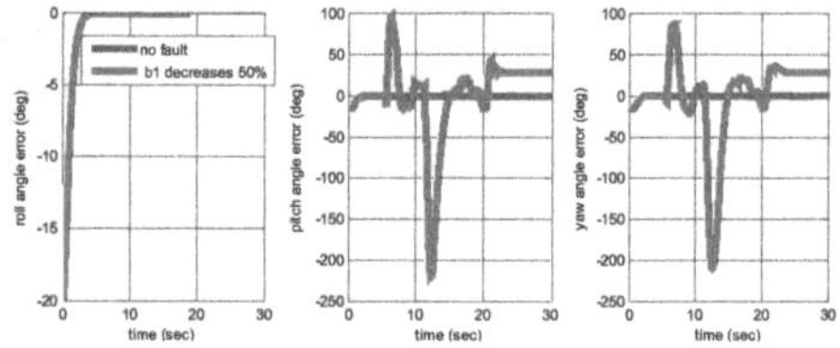

(b) Angle errors with $k_1 = 1/k_2 = 3$

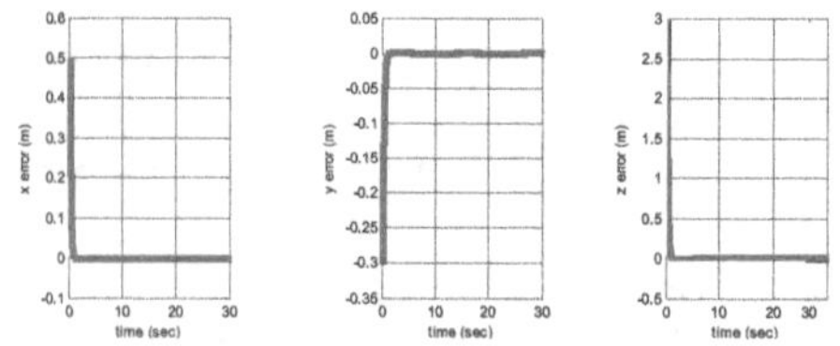

(c) Position errors with $k_1 = 5/k_2 = 30$

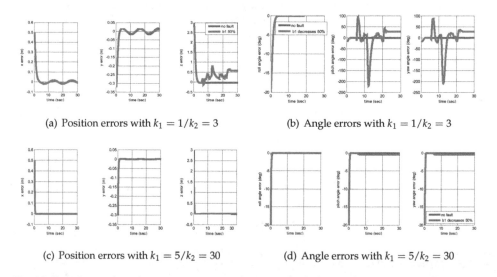

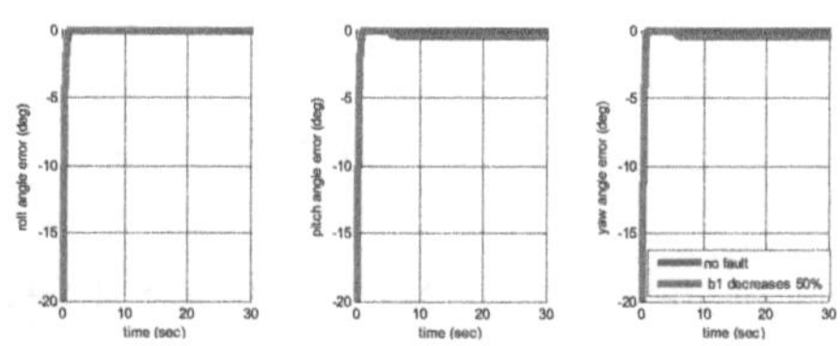

(d) Angle errors with $k_1 = 5/k_2 = 30$

Fig. 10. Position and angle error comparison between fault-free and 50% control effectiveness loss in first actuator, with control gains $k_1 = 1/k_2 = 3$ and $k_1 = 5/k_2 = 30$.

4.5 MPC simulation results

To illustrate the MPC approach, the quadrotor is assumed to be on the ground initially and it is required to reach a hover height of 4 m and stay in that height while stabilizing the pitch and roll angles. The x^d, y^d and ψ^d are $x^d = 2\ m$, $y^d = 3\ m$ and $\psi^d = 0$. The upper left plot of Figure 11 shows the time history of states for fault-free condition (the desired position values are dotted and the velocities are dashed lines).

The fault is designed to happen at time $t = 5$ s. At this time it is assumed that actuator faults occur which lead to multiple simultaneous partial loss of effectiveness of three actuators as follows: $\alpha_1 = 0.9$, $\alpha_2 = 0.7$, $\alpha_3 = 0.8$ and $\alpha_4 = 1.0$ (i.e. 10% loss of control effectiveness in the first motor, 30% in the second, 20% in the third and no fault in the fourth one). The upper right plot of Figure 11 shows the time history of the states when no fault estimation is performed. In fact in this case Algorithm 1 is used without any information about the occurred fault. One

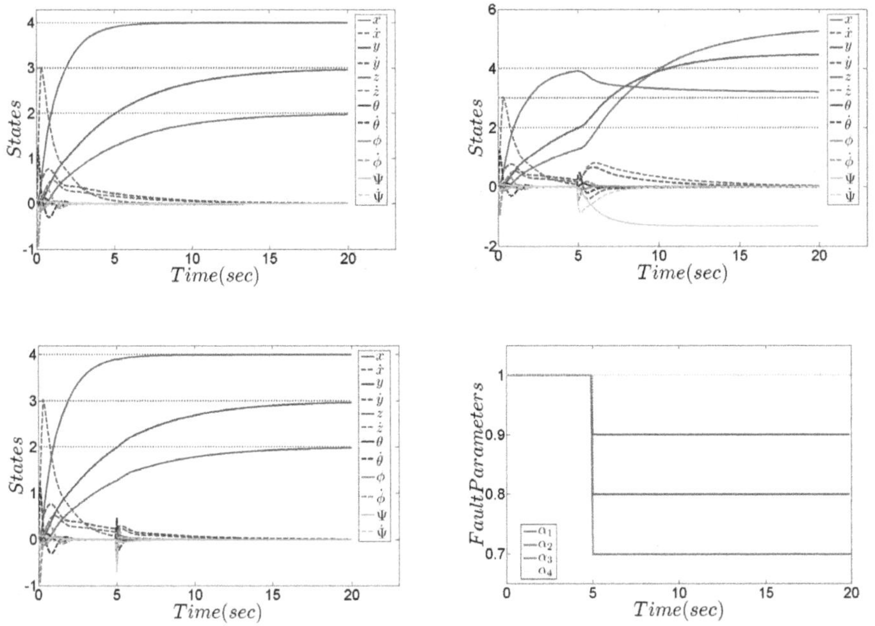

Fig. 11. MPC in fault-free and fault conditions without and with fault estimation.

can see that system states do not converge to their desired values and therefore it exhibits poor performance. The figure also shows that interestingly MPC exhibits some degree of fault tolerance inherently; all the linear and angular velocities are stabilized and there is only some error in the positions and orientations.

For the fault-tolerant MPC, Problem 2 is used where fault parameters estimation is provided by employing the Moving Horizon Estimator (MHE) Izadi et al. (2011b). The two lower plots of Figure 11 show fault parameters estimates as well as system states. It is clear that fault-tolerant MPC improves the system performance compared to the case without fault-tolerance.

4.6 FTPR simulation results

To illustrate the FTPR approach, let us assume that the quadrotor system is required to move from an initial position to a final one with $F_2(t_0) = 0$ and $F_3(t_0) = 0$, $F_2(t_f) = 20\ m$ and $F_3(t_f) = 30\ m$. It is also assumed that the maximal thrust that can be generated by each motor is $T_{max} = 8\ N$. ρ is set to 0.1 and the approach described is Section 3.6.3 gives the solution of $t_f = 6.81$ s. The simulation results for this scenario are given in Figure 12 which shows the system trajectories along the x and y directions. The figure also shows that the four applied thrusts do not exceed ρT_{max}.

For the fault scenario, it is assumed that the Qball-X4 loses 25% of the control effectiveness of its fourth rotor at the time instant $t = 2$ s. In this case and without trajectory re-planning, Figure 13(a) shows that the damaged system is not able anymore to reach the final desired

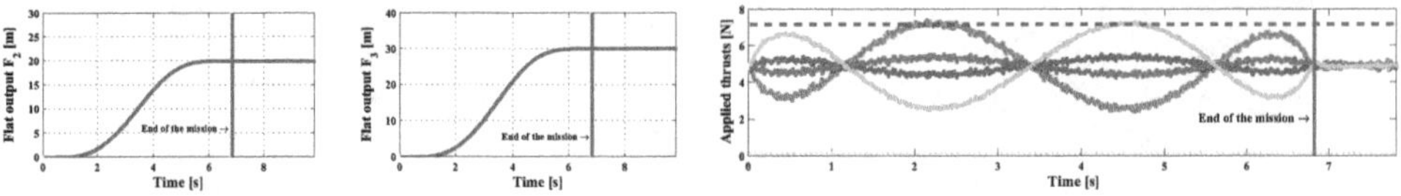

Fig. 12. System evolution in the x and y directions and the applied thrusts.

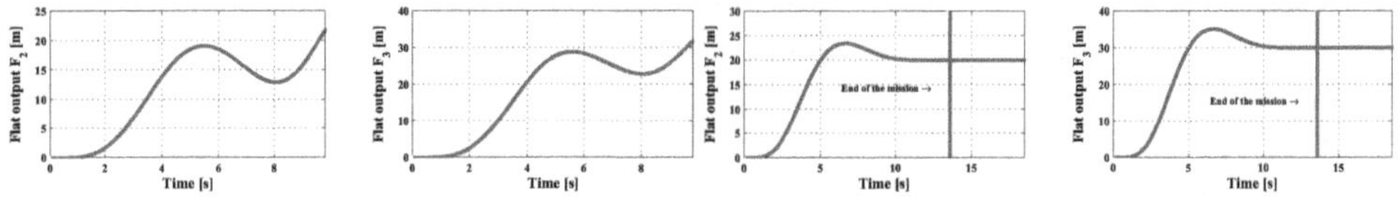

(a) Trajectories without re-planning

(b) Trajectories with re-planning

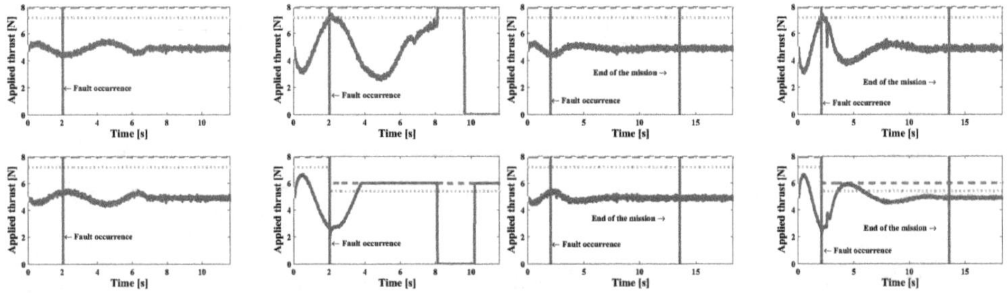

(c) Control inputs without re-planning

(d) Control inputs with re-planning

Fig. 13. Trajectories and control inputs without and with re-planning.

position. Figure 13(c) shows how the applied thrusts saturate when the system is forced to follow the pre-fault nominal trajectory. For the trajectory re-planning, once the fault is detected, isolated and identified, the trajectories are re-planned at $t = 2.5$ s where 0.5 s is the time assumed to be taken by the FDD module. The solution obtained from Section 3.6.4 is $t_f = 13.3$ s. Figure 13(b) shows that after re-planning, the UAV is able to reach the final desired position. The applied thrusts are illustrated in Figure 13(d) where it can be seen that the thrusts remain smaller than their maximal allowable limits. Thus, trajectory planning can help in keeping system stability by redefining the desired trajectories to be followed by taking into consideration the post-fault actuators limits.

5. Conclusion

This chapter presents some of the work that have been carried out at the Networked Autonomous Vehicles Laboratory of Concordia University. The main concern is to propose approaches that can be effective, easy to implement and to run on-board the UAVs. Many of the proposed methods have been implemented and tested with the Qball-X4 testbed and current work aims to propose and implement more advanced and practical techniques. As one of functional blocks in an AFTCS framework, FDD plays an important role for successful fault-tolerant control of systems (see Figure 3). Development on FDD techniques could not be included in the chapter due to space limit. Interested readers can refer to our recent work in

(Ma & Zhang, 2010a), (Ma & Zhang, 2010b), (Ma, 2011), (Gollu & Zhang, 2011), (Zhou, 2009) and (Amoozgar et al., 2011) for information on FDD techniques applied to UAV and satellite systems. Research work on fault-tolerant attitude control for spacecraft in the presence of actuator faults could not also be included but can be found in (Hu et al., 2011a), (Hu et al., 2011b), (Xiao et al., 2011a) and (Xiao et al., 2011b). Current and future work will focus more on the multi-vehicles cooperative control where preliminary works on cooperative control have been presented in (Izadi et al., 2009b), (Izadi et al., 2011a), (Sharifi et al., 2010), (Sharifi et al., 2011), (Mirzaei et al., 2011) and (Qu & Zhang, 2011).

6. Acknowledgements

Financial support of this work by the Natural Sciences and Engineering Research Council of Canada (NSERC) through a Strategic Project Grant (STPGP 350889-07) and Discovery Project Grants is highly acknowledged. Support from Quanser Inc. and colleagues from Quanser Inc. for the development of the Qball-X4 UAV testbed is also highly appreciated. Acknowledgements go also to the colleagues at the Department of Mechanical and Industrial Engineering of Concordia University, DRDC and other academic and industrial partners of the above grants, and all Postdoctoral Fellows/Research Associates, PhD, MASc/MEng students who contributed to this work.

7. References

Amoozgar, M. H., Gollu, N., Zhang, Y. M., Lee, J. & Ng, A. (2011). Fault detection and diagnosis of attitude control system for the JC2Sat-FF mission, *The 4th International Conference on Spacecraft Formation Flying Missions and Technology*, Saint Hubert, QC, Canada.

Bani Milhim, A. (2010a). *Modeling and Fault Tolerant PID Control of a Quad-rotor UAV*, Master's thesis, Concordia University, Montreal, QC, Canada.

Bani Milhim, A., Zhang, Y. M. & Rabbath, C. A. (2010b). Gain scheduling based PID controller for fault tolerant control of a quad-rotor UAV, *AIAA Infotech@Aerospace*, Atlanta, Georgia, USA.

Bresciani, T. (2008). *Modelling, Identification and Control of a Quadrotor Helicopter*, Master's thesis, Lund University, Sweden.

Chamseddine, A., Zhang, Y. M., Rabbath, C. A., Fulford, C. & Apkarian, J. (2011a). Model reference adaptive fault tolerant control of a quadrotor UAV, *AIAA Infotech@Aerospace*, St. Louis, Missouri, USA.

Chamseddine, A., Zhang, Y. M., Rabbath, C. A., Join, C. & Theilliol, D. (2011b). Flatness-based trajectory planning/re-planning for a quadrotor unmanned aerial vehicle, *IEEE Transactions on Aerospace and Electronic Systems (To appear)* .

Chen, H. & Allgower, F. (1998). A quasi infinitive horizon nonlinear model predictive control scheme with guaranteed stability, *Automatica*, 34(10): 1205–1217.

Dierks, T. & Jagannathan, S. (2008). Neural network output feedback control of a quadrotor UAV, *Proceedings of the 47th IEEE Conference on Decision and Control*, Cancun, Mexico, pp. 3633–3639.

Dierks, T. & Jagannathan, S. (2009). Neural network control of quadrotor UAV formations, *American Control Conference*, St. Louis, Missouri, USA, pp. 2990–2996.

Edwards, C., Lombaerts, T. & Smaili, H. (2010). *Fault Tolerant Flight Control: A Benchmark Challenge*, Lecture Notes in Control and Information Sciences, Springer.

Gollu, N. & Zhang, Y. M. (2011). Fault detection and diagnosis of a thruster controlled satellite for autonomous rendezvous and docking, *The 4th International Conference on Spacecraft Formation Flying Missions and Technology*, Saint Hubert, QC, Canada.

Haugen, F. (2004). *PID Control*, Tapir Academic Press.

Hu, Q.-L., Xiao, B. & Zhang, Y. M. (2011a). Fault-tolerant attitude control for spacecraft under loss of actuator effectiveness, *Journal of Guidance, Control, and Dynamics*, 3(34): 927–932.

Hu, Q.-L., Zhang, Y. M., Huo, X. & Xiao, B. (2011b). Adaptive integral-type sliding mode control for spacecraft attitude maneuvering under actuator stuck failures, *Chinese Journal of Aeronautics*, 1(24): 32–45.

Ioannou, P. A. & Sun, J. (1995). *Robust Adaptive Control*, Prentice Hall PTR.

Izadi, H. A. (2009a). *Decentralized Receding Horizon Control of Cooperative Vehicles with Communication Delays*, PhD thesis, Concordia University, Montreal, QC, Canada.

Izadi, H. A., Gordon, B. W. & Zhang, Y. M. (2009b). Decentralized receding horizon control for cooperative multiple vehicles subject to communication delay, *AIAA Journal of Guidance, Control, and Dynamics*, 6(32): 1959–1965.

Izadi, H. A., Gordon, B. W. & Zhang, Y. M. (2011a). Decentralized model predictive control for cooperative multiple vehicles subject to communication loss, *Special Issue on "Formation Flight Control" in International Journal of Aerospace Engineering*, 2011.

Izadi, H. A., Zhang, Y. M. & Gordon, B. W. (2011b). Fault tolerant model predictive control of quad-rotor helicopters with actuator fault estimation, *Proceedings of the 18th IFAC World Congress*, Milano, Italy.

Johnson, E. N., Chowdhary, G. V. & Kimbrell, M. S. (2010). Guidance and control of an airplane under severe structural damage, *AIAA Infotech@Aerospace*, Atlanta, Georgia, USA.

Krstic, M., Kanellakopoulos, I. & Kokotovic, P. V. (1995). *Nonlinear and Adaptive Control Design*, Wiley-Interscience, New York.

Li, T. (2011a). *Nonlinear and Fault-tolerant Control Techniques for a Quadrotor Unmanned Aerial Vehicle*, Master's thesis, Concordia University, Montreal, QC, Canada.

Li, T., Zhang, Y. M. & Gordon, B. (2011b). Fault tolerant control applied to a quadrotor unmanned helicopter, *Proceedings of the 7th ASME/IEEE International Conference on Mechatronics & Embedded Systems & Applications*, Washington DC, USA.

Ma, L. (2011). *Development of Fault Detection and Diagnosis Techniques with Applications to Fixed-wing and Rotary-wing UAVs*, Master's thesis, Concordia University, Montreal, QC, Canada.

Ma, L. & Zhang, Y. M. (2010a). Fault detection and diagnosis for GTM UAV with dual unscented Kalman filter, *AIAA Guidance, Navigation, and Control Conference*, Toronto, Ontario, Canada.

Ma, L. & Zhang, Y. M. (2010b). DUKF-based fault detection and diagnosis for GTM UAV using nonlinear and LPV models, *Proceedings of the 6th ASME/IEEE International Conference on Mechatronic & Embedded Systems & Applications*, Qingdao, P. R. China.

Maciejowski, J. M. & Jones, C. N. (2003). MPC fault-tolerant flight control case study: flight 1862, *IFAC Symposium on Safeprocess*, Washington D.C., USA.

MAST (2011). `https://www.grasp.upenn.edu/research/micro_autonomous_system_technologies_mast`.

Mirzaei, M., Sharifi, F., Gordon, B. W., Rabbath, C. A. & Zhang, Y. M. (2011). Cooperative multi-vehicle search and coverage problem in an uncertain environment, *Accepted by the 50th IEEE Conference on Decision and Control and European Control Conference*, Orlando, Florida, USA.

NAV (2011). http://users.encs.concordia.ca/~ymzhang/UAVs.htm.

Nguyen, H. V., Berbra, C., Lesecq, S., Gentil, S., Barraud, A. & Godin, C. (2009). Diagnosis of an inertial measurement unit based on set membership estimation, *The 17th Mediterranean Conference on Control and Automation*, Thessaloniki, Greece, pp. 211–216.

Noura, H., Theilliol, D., Ponsart, J. C. & Chamseddine, A. (2009). *Fault-tolerant Control Systems: Design and Practical Applications*, Springer.

Qu, Y. H. & Zhang, Y. M. (2011). Cooperative localization against GPS signal loss in multiple UAVs flight, *Invited Special Issue on "Fault Detection, Diagnosis and Tolerant Control" in Journal of Systems Engineering and Electronics*, 1(22): 103–122.

Quanser (2010). Quanser Qball-X4 User Manual. Document number 829.

Rafaralahy, H., Richard, E., Boutayeb, M. & Zasadzinski, M. (2008). Simultaneous observer based sensor diagnosis and speed estimation of unmanned aerial vehicle, *Proceedings of the 47th IEEE Conference on Decision and Control*, Cancun, Mexico, pp. 2938–2943.

Sadeghzadeh, I., Mehta, A., Zhang, Y. M. & Rabbath, C. A. (2011). Fault-tolerant trajectory tracking control of a quadrotor helicopter using gain-scheduled PID and model reference adaptive control, *Annual Conference of the Prognostics and Health Management Society*, Montreal, QC, Canada.

Sharifi, F., Gordon, B. W. & Zhang, Y. M. (2010). Decentralized sliding control of cooperative multi-agent systems subject to communication delays, *AIAA Guidance, Navigation, and Control Conference*, Toronto, Ontario, Canada.

Sharifi, F., Zhang, Y. M. & Gordon, B. W. (2011). Voroni-based coverage control for multi-quadrotor UAVs, *Proceedings of the 7th ASME/IEEE International Conference on Mechatronic & Embedded Systems & Applications*, Washington, DC, USA.

Singh, G. K. & Holé, K. E. (2004). Guaranteed performance in reaching mode of sliding mode controlled systems, *Sadhana*, 29(1): 129–141.

STARMAC (2011). http://hybrid.stanford.edu/~starmac/project.htm.

Stepanyan, V. & Krinshnakumar, K. (2010a). Input and output performance of M-MRAC in the presence of bounded disturbances, *AIAA Guidance, Navigation, and Control Conference*, Toronto, Ontario, Canada.

Stepanyan, V. & Krinshnakumar, K. (2010b). MRAC revisited: guaranteed performance with reference model modification, *American Control Conference*, Baltimore, Maryland, USA, pp. 93–98.

SWARM (2011). http://vertol.mit.edu.

Xiao, B., Hu, Q.-L. & Zhang, Y. M. (2011a). Fault-tolerant attitude control for flexible spacecraft without angular velocity magnitude measurement, *Journal of Guidance, Control, and Dynamics*, 5(34): 1556–1561.

Xiao, B., Hu, Q.-L. & Zhang, Y. M. (2011b). Adaptive sliding mode fault tolerant attitude tracking control for flexible spacecraft under actuator saturation, *IEEE Transactions on Control Systems Technology (Available on-line on 25 October 2011)* .

Zhang, X. (2010a). *Lyapunov-based Fault Tolerant Control of Quadrotor Unmanned Aerial Vehicles*, Master's thesis, Concordia University, Montreal, QC, Canada.

Zhang, X., Zhang, Y. M., Su, C.-Y. & Feng, Y. (2010b). Fault tolerant control for quadrotor via backstepping approach, *The 48th AIAA Aerospace Sciences Meeting Including the New Horizons Forum and Aerospace Exposition*, Orlando, Florida, USA.

Zhang, Y. M. & Jiang, J. (2008). Bibliographical review on reconfigurable fault-tolerant control systems, *IFAC Annual Reviews in Control*, 32(2): 229–252.

Zhou, Q.-L. (2009). *Reconfigurable Control Allocation Design with Applications to Unmanned Aerial Vehicle and Aircraft*, Master's thesis, Concordia University, Montreal, QC, Canada.

Adaptive Feedforward Control for Gust Loads Alleviation

Jie Zeng[1], Raymond De Callafon[2] and Martin J. Brenner[3]
[1]*ZONA Technology, Inc.*
[2]*University of California, San Diego*
[3]*NASA Dryden Flight Research Center*
USA

1. Introduction

Active control techniques for the gust loads alleviation/flutter suppression have been investigated extensively in the last decades to control the aeroelastic response, and improve the handling qualities of the aircraft. Nonadaptive feedback control algorithms such as classical single input single output techniques (Schmidt & Chen, 1986), linear quadratic regulator (LQR) theory (Mahesh et al., 1981; Newsom, 1979), eigenspace techniques (Garrard & Liebst, 1985; Leibst et al., 1988), optimal control algorithm (Woods-Vedeler et al., 1995), H_∞ robust control synthesis technique (Barker et al., 1999) are efficient methods for the gust loads alleviation/flutter suppression. However, because of the time varying characteristics of the aircraft dynamics due to the varying configurations and operational parameters, such as fuel consumption, air density, velocity, air turbulence, it is difficult to synthesize a unique control law to work effectively throughout the whole flight envelope. Therefore, a gain scheduling technique is necessary to account for the time varying aircraft dynamics.

An alternative methodology is the feedforward and/or feedback adaptive control algorithms by which the control law can be updated at every time step (Andrighettoni & Mantegazza, 1998; Eversman & Roy, 1996; Wildschek et al., 2006). With the novel development of the airborne LIght Detection and Ranging (LIDAR) turbulence sensor available for an accurate vertical gust velocity measurement at a considerable distance ahead of the aircraft (Schmitt, Pistner, Zeller, Diehl & Navé, 2007), it becomes feasible to design an adaptive feedforward control to alleviate the structural loads induced by any turbulence and extend the life of the structure. The adaptive feedforward control algorithm developed in (Wildschek et al., 2006) showed promising results for vibration suppression of the first wing bending mode. However, an unavoidable constraint for the application of this methodology is the usage of a high order Finite Impulse Response (FIR) filter. As a result, an overwhelming computation effort was needed to suppress the structural vibration of the aircraft.

In this chapter, an adaptive feedforward control algorithm where the feedforward filter is parameterized using orthonormal basis expansions along with a recursive least square algorithm with a variable forgetting factor is proposed for the feedforward compensation of gust loads. With the use of the orthonormal basis expansion, the prior flexible modes information of the aircraft dynamics can be incorporated to build the structure of the feedforward controller. With this strategy, the order of the feedforward filter to be estimated can be largely reduced. As a result, the computation effort is greatly decreased, and the performance of the feedforward controller for gust loads alleviation will be enhanced. Furthermore, an FFT based PolyMAX identification method and the stabilization diagram program (Baldelli et al., 2009) are proposed to estimate the flexible modes of the aircraft dynamics.

The need for an integrated model of flight dynamics and aeroelasticity is brought about by the emerging design requirements for slender, more flexible and/or sizable aircraft such as the Oblique Flying Wing (OFW), HALE, Sensorcraft and morphing vehicles, etc. Furthermore, a desirable unified nonlinear simulator should be formulated in principle by using commonly agreeable terms from both the flight dynamics and aeroelasticity fields in a consistent manner.

A unified integration framework that blends flight dynamics and aeroelastic modeling approaches with wind-tunnel or flight-test data derived aerodynamic models has been developed in (Baldelli & Zeng, 2007). This framework considers innovative model updating techniques to upgrade the aerodynamic model with data coming from CFD/wind-tunnel tests for a rigid configuration or data estimated from actual flight tests when flexible configurations are considered.

Closely following the unified integration framework developed in (Baldelli & Zeng, 2007), an F/A-18 Active Aeroelastic Wing (AAW) aeroelastic model with gust perturbation is developed in this chapter, and this F/A-18 AAW aeroelastic model can be implemented as a test-bed for flight control system evaluation and/or feedback/feedforward controller design for gust loads alleviation/flutter suppression of the flexible aircraft.

The outline of the chapter is as follows. In Section II, a feedforward compensation framework is introduced. Section III presents the formulation of the orthonormal finite impulse filter structure. A brief description of a frequency domain PolyMAX identification method is presented in Section IV. In Section V, a recursive least square estimation method with variable forgetting factor is discussed. Section VI includes the development of a linear F/A-18AAW aeroelastic model and the application of the adaptive feedforward controller to F/A-18 AAW aeroelastic model.

2. Basic idea of the feedforward controller

In order to analyze the design of the feedforward controller, F, consider the simplified block diagram of structural vibration control of a single input signal output (SISO) dynamic system depicted in Fig. 1. The gust perturbation, $w(t)$, passes through the primary path, H, the body of the aircraft, to cause the structural vibrations. Mathematically, H can be characterized as the model/transfer function from the gust perturbation to the accelerometer sensor position. The gust perturbation, $w(t)$, can be measured by the coherent LIDAR beam airborne wind sensor. The measured signal, $n(t)$, is fed into the feedforward controller, F, to calculate the control surface demand, $u(t)$, for vibration compensations. The structural vibrations are measured by the accelerometers providing the error signal, $e(t)$. G is the model/transfer function from the corresponding control surface to the accelerometer sensor position, and which is so called the secondary path.

In order to apply the feedforward control algorithm for gust loads alleviation, developing a proper sensor to accurately measure the gust perturbation is crucial for the success of the feedforward control application. As mentioned in (Schmitt, Pistner, Zeller, Diehl & Navé, 2007), such a sensor should meet the following criteria:

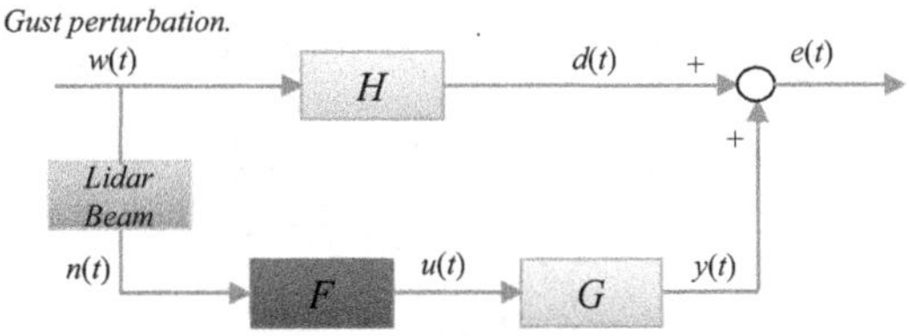

Fig. 1. Block Diagram of the Structural Vibration Control with Feedforward Compensation.

- A feedforward-looking measuring of 50 to 150 m to ensure that the measured air flow is the one actually affecting the aerodynamics around the aircraft.
- The sensor must be able to measure the vertical wind speed.
- The standard deviation of the wind speed measurement should be small, at least in the range of [2-4] m/s.
- The sensor must be able to produce reliable signals in the absence of aerosols.
- A good longitudinal resolution (the thickness of the air slice measured ahead).
- A good temporal resolution.

A sensor system that meets these requirements is a so-called short pulse UV Doppler LIDAR, and was developed in (Schmitt, Pistner, Zeller, Diehl & Navé, 2007). This short pulse UV Doppler LIDAR was successfully applied to an Airbus 340 to measure the vertical gust speed (Schmitt, Rehm, Pistner, Diehl, Jenaro-Rabadan, Mirand & Reymond, 2007). The authors in (Schmitt, Rehm, Pistner, Diehl, Jenaro-Rabadan, Mirand & Reymond, 2007) claimed that the system is ready to be used to design feedforward control for gust loads alleviation.

Assuming a perfect gust perturbation signal can be measured via the LIDAR beam sensor, that means, $n(t) = w(t)$, the error signal, $e(t)$, can be described by

$$e(t) = [H(q) + G(q)F(q)]\, w(t) \tag{1}$$

In case the transfer functions in Eq. (1) are known, an ideal feedforward controller, $F(q) = F_i(q)$, can be obtained by

$$F_i(q) = -\frac{H(q)}{G(q)} \tag{2}$$

in case, $F_i(q)$, is a stable and causal transfer function. The solution of $F_i(q)$ in Eq. (2) assumes full knowledge of $G(q)$ and $H(q)$. Moreover, the filter, $F_i(q)$, may not be a causal or stable filter due to the dynamics of $G(q)$ and $H(q)$ that dictate the solution of the feedforward controller, $F_i(q)$. An approximation of the feedforward filter, $F_i(q)$, can be made by an output-error based optimization that aims at finding the best causal and stable approximation, $F(q)$, of the ideal feedforward controller in $F_i(q)$ in Eq. (2).

A direct adaptation of the feedforward controller $F(q,\theta)$ can be performed by considering the parameterized error signal, $e(t,\theta)$,

$$e(t,\theta) = H(q)w(t) + F(q,\theta)G(q)w(t). \tag{3}$$

Defining the signals

$$d(t) := H(q)w(t),\ u_f(t) := -G(q)w(t) \tag{4}$$

where $d(t)$ can actually be measured, and $u_f(t)$ is called filtered input signal, Eq. (3) is reduced to

$$e(t,\theta) = d(t) - F(q,\theta)u_f(t) \tag{5}$$

for which the minimization

$$\min_{\theta} \frac{1}{N} \sum_{t=1}^{N} e^2(t,\theta) \tag{6}$$

to compute the optimal feedforward filter, $F(q,\theta)$, is a standard output-error (OE) minimization problem in a prediction error framework (Ljung, 1999).

The minimization of Eq. (6) for $\lim_{N\to\infty}$ can be rewritten into the frequency domain expression

$$\min_{\theta} \int_{\pi}^{-\pi} |H(e^{jw}) + G(e^{jw})F(e^{jw},\theta)|^2 \tag{7}$$

using Parceval's theorem (Ljung, 1999). It can be observed that the standard output-error (OE) minimization problem in Eq. (6) can be used to compute the optimal feedforward filter $F(q,\theta)$, provided $d(t)$ and $u_f(t)$ in Eq. (4) are available.

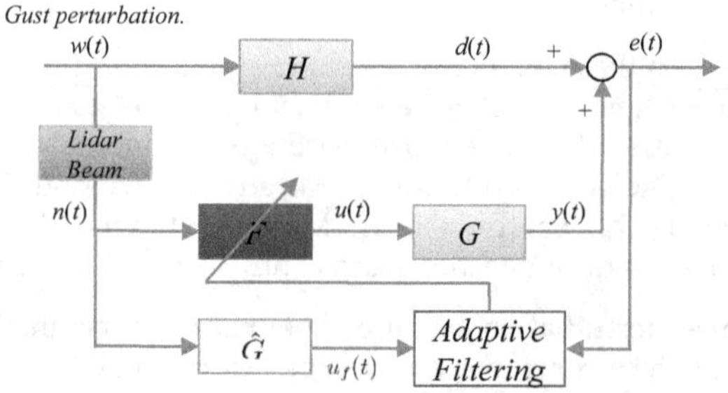

Fig. 2. Block Diagram of the Structural Vibration Control with Adaptive Feedforward Compensation.

For a proper derivation of the adaptation of the feedforward filter, F, an approximate model, $\hat{G}$, of the control path, G, is required to create the filtered signal, $u_f(t)$, for adaptive filtering purpose. The adaptation of the feedforward filter is illustrated in Fig. 2. The filtered signal, $u_f(t)$, and the error signal, $e(t)$, are used for the computation of the coefficients of the feedforward filter by the adaptive filtering. Thus, the coefficients of the feedforward filter, F, can be updated at each time constant for structural vibration reduction.

3. ORThonormal Finite Impulse Response (ORTFIR) filter structure

In general, the feedforward filter, F, in Fig. 1 can be realized by adopting both the finite impulse response (FIR) structure as well as the infinite impulse response (IIR) structure. Because the FIR filter incorporates only zeros, it is always stable and it will provide a linear phase response. It is the most popular adaptive filter widely used in adaptive filtering.

Generally, the discrete time linear time invariant (LTI) FIR filter, $F(q)$, can be presented as:

$$F(q) = \sum_{k=0}^{L-1} \beta_k q^{-k} \tag{8}$$

where q^{-1} denotes the usual time shift operator, $q^{-1}x(t) = x(t-1)$. Adaptive filter estimation using FIR filters converges to optimal and unbiased estimates irrespective of the coloring of the noise on the output data. However, a FIR filter is usually too simple to model complex system dynamics such as AE/ASE systems with many resonance modes being excited by atmospheric perturbations. As a result, many tapped delay coefficients of the FIR filter are required to approximate the optimal filter. Even though an IIR filter is appealing as an alternative, the inherent stability and bias estimation problems limit the use of an IIR filter for adaptive filtering in aeroservoelastic systems.

To improve the approximation properties of the adaptive filter, F, in Fig. 1, the linear combination of tapped delay functions, q^{-1}, in the FIR filter of Eq. (8) can be generalized to the following form:

$$F(q,\theta) = \sum_{k=0}^{L-1} \beta_k B_k(q) \tag{9}$$

where $B_k(q)$ are generalized (orthonormal) basis functions (Heuberger et al., 1995) that contain some *a-priori* knowledge on the desired filter dynamics. In other words, the orthonormal basis functions that are used in the parametrization of the ORTFIR filter will be tuned on the fly by taking full advantage of the modal information embedded in the flight data.

Construction of the Orthonormal Basis Sets

The application of orthonormal basis functions to parameterize and estimate dynamical systems has obtained extensive attention in recent years. Different constructions of the orthonormal basis structure has been reported in (Heuberger et al., 1995; Ninness & Gustafsson, 1997; Zeng & de Callafon, 2006). It is assumed that the pole locations are already known with the use of the standard open-loop prediction error system identification methods. Suppose the poles $\{\xi_i\}_{i=1,2,\cdots,N}$ are known, an all pass function, $P(q)$, can be created by these poles, and is given as

$$P(q) = \prod_{i=1}^{N} \left[\frac{1-\xi_i^* q}{q-\xi_i} \right] \tag{10}$$

Let (A,B,C,D) be a minimal balanced realization of an all pass function, $P(q)$, define the input to state transfer function, $B_0(q) = (qI-A)^{-1}B$, then a set of functions, $B_i(q)$, can be obtained via

$$B_i(q) = B_0(q)P^i(q) \tag{11}$$

and $B_i(q)$ has orthogonal property

$$\frac{1}{2\pi j}\oint B_i(q)B_k^T(1/q)\frac{dq}{q} = \begin{cases} I & i=k \\ 0 & i\neq k \end{cases} \tag{12}$$

The construction of the orthonormal basis function is illustrated in Fig. 3. It should be noted that if $B_0(q)$, includes all information of a dynamical system, then only one parameter, β_0, needs to be estimated to approximate this dynamic system. It means that the parameters estimated will directly depend on the *a-priori* system information injected into the basis functions, $B_i(q)$. An important property and advantage of the ORTFIR filter is the knowledge

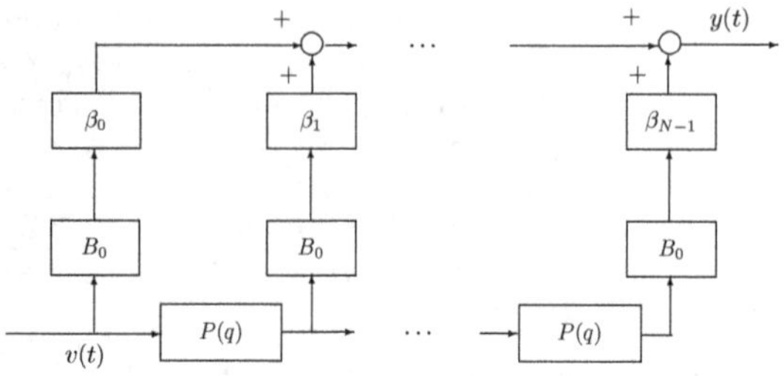

Fig. 3. ORTFIR Filter Topology.

of the (desired) dynamical behavior can be incorporated throughout the basis functions, $B_i(q)$. As a result, an accurate description of the filter to be estimated can be achieved by a relatively small number of coefficients.

Case Example: Illustration of the Advantage of Using ORTFIR Filter Over FIR Filter

A 4 degrees-of-freedom (DOF) lumped parameter system is considered to demonstrate the advantage of using ORTFIR filter over FIR filter. An illustration of this 4-DOF lumped parameter system is shown in Fig. 4, where k_i and $c_i (i = 1, \cdots, 5)$ indicate the system stiffness and damping, respectively and $m_i (i = 1, \cdots, 4)$ are the masses. The nominal values of these parameters are given as,

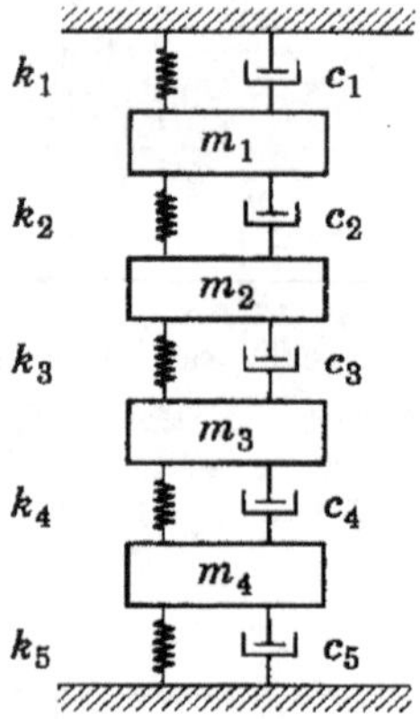

Fig. 4. Lumped Parameter System.

$$
\begin{aligned}
&m_1 = m_2 = m_3 = m_4 = 1 \\
&k_1 = k_3 = k_5 = 1750 \\
&k_2 = k_4 = 2000 \\
&c_1 = c_3 = c_5 = 0.7 \\
&c_2 = c_4 = 0.8
\end{aligned}
\tag{13}
$$

A mathematical model of this lumped system can be easily derived with the use of Newton's second law. The natural frequencies and damping ratios of this lumped system are also obtained. For simplicity purposes, all the units of this 4-DOF lumped system are omitted. This mathematical model is applied in this case example as the real model, and an FIR model and an ORTFIR model will be implemented to approximate this real model, respectively. To facilitate the model estimation using input and output data of the 4-DOF lumped parameter system, a band limited white noise (zero mean) is injected to the 4-DOF lumped parameter system, and an additional band limited white noise (zero mean) is added to the output response to simulate the measurement noise. With the collected input/output data, an FIR filter with varying order is applied to fit the real model, the variance of the simulation error (the difference of the measured and the simulated output) is used to indicate the performance of the FIR filter. Furthermore, the PolyMAX identification method described in Section 4 is applied in this case example to estimate the four physical modes. These estimated physical modes (shown in Table 2, Section 4) are used for the basis function generation of the ORTFIR filter. Finally, an eight order ORTFIR filter is applied to approximate the physical system. The estimation results are shown in Table 1. From Table 1, it is clearly seen that with FIR filter, the optimal FIR filter will be 400th order, with the smallest simulation error 36.18. However, with the simplest eight order ORTFIR filter, the variance of the simulation error is only 13.18, which is almost three times smaller than that of 400th order FIR filter. Fig. 5 compared the model estimation results using 50th/400th order FIR filters and 8th order ORTFIR filter. From Fig. 5, it is observed that with 50th order FIR filter, the essential dynamics of the physical system can hardly be catched. With 400th order FIR filter, even though the physical system can be correctly approximated, the estimated model has evident variation, especially in the high frequency range. On the other hand, with 8th order ORTFIR filter, the physical system can be perfectly approximated, no visible variation of the estimated model was found in a wider frequency range.

Order	1000	500	400	300	200	100	50	8
Filter Type	FIR	FIR	FIR	FIR	FIR	FIR	FIR	ORTFIR
Variance of Simulation Error	42.53	36.46	36.18	36.70	40.28	52.91	70.10	13.18

Table 1. Model Estimation Results Using FIR Fiter and ORTFIR Filter.

4. Modal parameters estimation-frequency PolyMAX identification

A rather general frequency-domain identification method using the standard least squares estimator algorithm is introduced and applied to extract the modal characteristics of a dynamic system from a set of measured data. Consider a set of noisy complex Frequency Response Functions (FRF) measurement data, $G(\omega_j)$, $(j = 1, \cdots, N)$. The approximation of the data by a model $P(\omega)$ is addressed by considering the following additive error,

$$E(\omega_j) = G(\omega_j) - P(\omega_j) \quad j = 1, \cdots, N \tag{14}$$

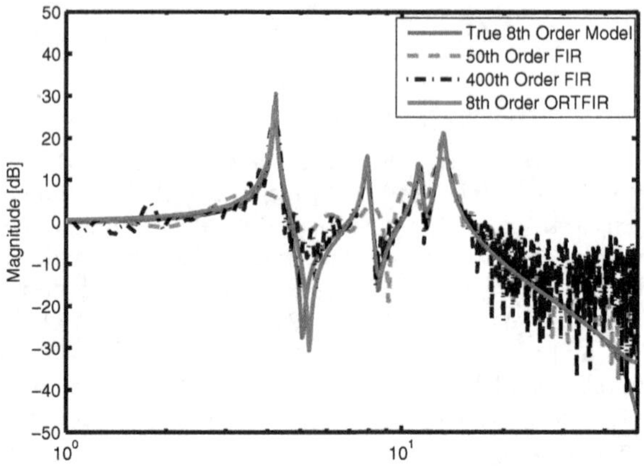

Fig. 5. Comparison of the Model Estimation Using the FIR and ORTFIR Filters.

Then, it is assumed that the model, $P(\omega)$, can be represented by a right polynomial fraction matrix given by,

$$P(\omega) = [B(\omega)][A(\omega)]^{-1} \tag{15}$$

where $P(\omega) \in \mathcal{C}^{p\times m}$ is the FRF matrix with p outputs and m inputs, $B(\omega) \in \mathcal{C}^{p\times m}$ is the numerator matrix polynomial, and $A(\omega) \in \mathcal{C}^{m\times m}$ is the denominator matrix polynomial.

The matrix polynomial $B(\omega)$ is parameterized by

$$B(\omega) = \sum_{k=0}^{n_b} B_k \xi_k(\omega) \tag{16}$$

where $B_k \in \mathcal{R}^{p\times m}$, and n_b is the number of non-zero matrix coefficients in $B(\omega)$, or the order of $B(\omega)$. $\xi_k(\omega)$ are the polynomial basis functions. For continuous time model, $\xi_k(\omega) = -i\omega_k$. For discrete time model, $\xi_k(\omega) = e^{-i\omega_k T}$ (T is the sampling time).

The matrix polynomial $A(\omega)$ is parameterized by

$$A(\omega) = \sum_{k=0}^{n_a} A_k \xi_k(\omega) \tag{17}$$

where $A_k \in \mathcal{R}^{m\times m}$, and n_a is the number of non-zero matrix coefficients in $A(\omega)$.

Assuming that the coefficients of the denominator, $A(\omega)$, are $[A_0, A_1, \cdots, A_{n_a}]$, then, a constraint, $A_0 = I_m$, is set to obtain a stable model to fit the measured frequency-domain data. Here, a constraint, $A_{n_a} = I_m$, is adopted to extract physical modes from the measured frequency-domain data (Cauberghe et al., 2004).

With, $A_{n_a} = I_m$, the poles of the estimated model are separated into *stable physical* poles and *unstable mathematical* poles, from which a very clean stabilization diagram can be obtained, and the *physical* modal parameters of the real system can be estimated from a quick evaluation of the generated stabilization diagram (Cauberghe et al., 2005).

The stabilization diagram assumes an increasing model order (number of poles noted in the left ordinate axis), and it indicates where on the frequency axis the poles are located. As a

$$\xi(t) = y(t) - \hat{\theta}^T(t-1)\phi(t) \tag{20}$$

$$k(t) = \frac{P(t-1)\phi(t)}{\lambda_1(t) + \phi^T(t)P(t-1)\phi(t)} \tag{21}$$

2. Update the inverse correlation matrix, $P(t)$, and the forgetting factor, $\lambda(t)$:

$$P(t) = \lambda(t)^{-1}\left[1 - k(t)\phi^T(t)\right]P(t-1) \tag{22}$$

$$\begin{aligned} \lambda(t) &= \lambda_{min} + (1-\lambda_{min}) \cdot 2^{-L(t)} \\ L(t) &= \quad round(\rho \cdot \xi(t)^2) \end{aligned} \tag{23}$$

where ρ is a design parameter which controls the change rate and the width of a unity zone, $\xi(t)$ is the estimation error which is calculated via Eq. (20). $L(t)$ is defines as the nearest integer of $\rho \cdot \xi(t)^2$ at each instant time step. λ_{min} defines the lower bound of the λ.

In Eq. (23), it is shown that when the estimation error, $\xi(t)$ and $L(t)$ is small, $2^{-L(t)} \longrightarrow 1$, and $\lambda(t) \longrightarrow 1$ at an exponential rate, and this rate is controller by ρ. When $\xi(t)$ increases to infinity, λ reaches its minimum value. The RLS minimization is posted as:

$$J(t) = \sum_{i=1}^{t} \lambda(i)^{t-i}[y(i) - \hat{\theta}(t)^T\phi(i)]^2 \tag{24}$$

By choosing the variable forgetting factor indicated in Eq. (23), the fast decrease of the inverse correlation matrix, $P(t)$, can be avoided at the beginning of the estimation. In general this will result in an accelerated convergence by maintaining a high adaptation at the beginning of the estimation when the parameters, θ, are still far from the optimal value.

6. Application to closed loop F/A-18 AAW linear model

6.1 Linear aeroelastic solver formulation approach

A unified aeroelastic formulation to take into account the influence of aeroelastic effects on the flight dynamic behavior of the whole aircraft has been developed in (Baldelli et al., 2006).

A general formulation of a flexible aircraft with respect to a body-fixed reference system driven by aerodynamic, thrust, and gravity (g) forces and moments can be defined as:

$$m\left[\dot{V}_b + \Omega_b \times V_b - R_{bg}(E)\,[0,\,0,\,g]^T\right] = F_A + F_\delta + F_T + \Delta_F$$

$$J\dot{\Omega}_b + \Omega_b \times J\Omega_b = M_A + M_\delta + M_T + \Delta_M \tag{25}$$

where, m and J are the air vehicle mass and inertia tensor, and $R_{bg}(E)$ is the rotation mapping from inertial to body-axes, ($E = [\,\phi,\ \theta,\ \psi\,]$).

Eq. (25) is driven by the forces and moments on its right hand side, where F_A and M_A are the external aerodynamic forces and moments on the air vehicle. F_A and M_A are a function of the aerodynamic flight states ($V\,\alpha,\ \beta,\ E, \ldots,\ etc.$), Mach number, body angular rates (Ω_b), and control surface deflections and are usually obtained by wind-tunnel or flight tests. In either case, the quasi-steady influence of the deformed air-vehicle is included by considering flexible-to-rigid ratios or Parameter Identification (PID) techniques, (Morelli, 1995; Morelli & Klein, 1997). F_δ and M_δ are the aerodynamic forces and moments from the control surfaces commanded by the flight control system and pilot inputs while F_T and M_T includes the thrust loads.

In addition, Δ_F and Δ_M are the *aeroelastic incremental loads* due to the structural deformation. Usually, these loads are assumed to be quasi-statics and can be computed by a static aeroelastic analysis. However, this quasi-static assumption may not be sufficient for a highly reconfigurable and flexible aircraft like the new generation of Morphing UAV, HALEs, etc., where the interaction between the dynamic structural deformation due to unsteady flow and rigid body motion can play an important role.

During the integration process, the aeroelastic equations of motion underwent two similarity transformation steps, so the generalized coordinates related with the six rigid-body modes originally defined in principle axes are mapped into the airframe states (stability-axes definition). Specifically, for symmetric maneuvers the transformation matrix, $[T_A]_{long}$ (Baldelli et al., 2006), reads as:

$$\begin{Bmatrix} T_x \\ T_z \\ R_y \\ \dot{T}_x \\ \dot{T}_z \\ \dot{R}_y \end{Bmatrix} = \underbrace{\begin{bmatrix} -1 & 0 & 0 & 0 & 0 & 0 \\ 0 & 0 & 1 & 0 & 0 & 0 \\ 0 & 0 & 0 & 0 & 1 & 0 \\ 0 & -1 & 0 & 0 & 0 & 0 \\ 0 & 0 & 0 & -V & V & 0 \\ 0 & 0 & 0 & 0 & 0 & 1 \end{bmatrix}}_{[T_A]_{long}} \begin{Bmatrix} x \\ u \\ h \\ \alpha \\ \theta \\ q \end{Bmatrix} \tag{26}$$

For an anti-symmetric maneuver, $[T_A]_{lat}$ is,

$$\begin{Bmatrix} T_y \\ R_x \\ R_z \\ \dot{T}_y \\ \dot{R}_x \\ \dot{R}_z \end{Bmatrix} = \underbrace{\begin{bmatrix} 1 & 0 & 0 & 0 & 0 & 0 \\ 0 & 0 & 0 & 0 & -1 & 0 \\ 0 & 0 & 0 & 0 & 0 & -1 \\ 0 & V & 0 & 0 & 0 & V \\ 0 & 0 & -1 & 0 & 0 & 0 \\ 0 & 0 & 0 & -1 & 0 & 0 \end{bmatrix}}_{[T_A]_{lat}} \begin{Bmatrix} y \\ \beta \\ p \\ r \\ \phi \\ \psi \end{Bmatrix} \tag{27}$$

For an asymmetric maneuver, the matrix $[T_A] \in \mathbf{R}^{12\times 12}$ will be composed by the proper allocation of the elements that form the rows and columns of the $[T_A]_{long}$ and $[T_A]_{lat}$ matrices.

In this new coordinate system, the linear aeroelastic Equations of Motion (EoM) are:

$$\left\{ \begin{bmatrix} M_{rr} & 0 \\ 0 & M_{ee} \end{bmatrix} s^2 + \begin{bmatrix} C_{rr} & 0 \\ 0 & C_{ee} \end{bmatrix} s + \begin{bmatrix} K_{rr} & 0 \\ 0 & K_{ee} \end{bmatrix} \right\} \begin{Bmatrix} \xi_{as} \\ \eta_e \end{Bmatrix} =$$

$$q_\infty \left\{ \begin{bmatrix} Q_{rr}(s) & Q_{re}(s) \\ Q_{er}(s) & Q_{ee}(s) \end{bmatrix} \begin{Bmatrix} \xi_{as} \\ \eta_e \end{Bmatrix} + \begin{bmatrix} Q_{r\delta}(s) \\ Q_{e\delta}(s) \end{bmatrix} \delta_u + \begin{bmatrix} \frac{1}{V} Q_{rG}(s) \\ \frac{1}{V} Q_{eG}(s) \end{bmatrix} w_G \right\} \tag{28}$$

where w_G is the gust input; the elastic generalized coordinates, η_e, input, and δ_u, vectors are,

$$\eta_e^T = \left[\eta_{e_1}, \ldots, \eta_{e_{Ne}}\right]^T$$
$$\delta_u^T = \left[\delta_{elev}, \delta_{ail}, \delta_{rud}, \ldots\right]^T$$

It should be noted that the equations are only coupled via external forces and moments. In addition, after the transformation is applied the generalized mass matrix of the finite element model, it is no longer necessarily diagonal. In fact, the sub-matrix, M_{rr}, associated with the rigid body modes is identical to the mass matrix in the flight dynamics equation; i.e. the off-diagonal terms contain the products of inertia,

$$M_{rr} = diag(mI_3, J) \tag{29}$$

Usually, the aerodynamic force coefficient matrix, $Q(s)$, is approximated using the Rational Function Approximation (RFA) approach as

$$Q(s) = [A_0] + [A_1]\frac{L}{V}s + [A_2]\frac{L^2}{V^2}s^2 + [D]\left(sI - \frac{V}{L}[R]\right)^{-1}[E]\,s \tag{30}$$

where the $[A_i]$, $i = 0,\ 1,\ 2$, $[D]$ and $[E]$ matrices are column partitioned as,

$$[A_i] = \begin{bmatrix} A_r & A_e & A_\delta \end{bmatrix} \tag{31}$$

where $i = r,\ e,\ \delta$ are the airframe, elastic and control related states.

In this formulation, the $[A_i]$ coefficient matrices represent the quasi-steady aerodynamic forces, and the remanent terms are used to model the flow unsteadiness by Padé approximation.

Using the Minimum-State approach during the RFA implemented in the `ZAERO/ASE` module (Karpel, 1992), and due to the performed similarity transformation the aero-lag terms are computed as,

$$\{\dot{x}_L\} = \frac{V}{L}[R]\{x_L\} + \begin{bmatrix} E_{L*} & E_{Lr} & E_{Le} & E_{L\delta} \end{bmatrix} \begin{Bmatrix} \xi_{as} \\ \dot{\xi}_{as} \\ \dot{\eta}_e \\ \delta_{\dot{u}} \end{Bmatrix} \tag{32}$$

$$z_L = \begin{bmatrix} D \end{bmatrix} \begin{Bmatrix} x_L \end{Bmatrix} \tag{33}$$

By including Eqs. (30) and (32) into Eq. (28), the aeroelastic EoM can now be easily partitioned in accordance with the airframe degrees of freedom, elastic dynamics, aerodynamic lag terms, a set of control inputs and gust perturbation as:

$$\begin{Bmatrix} \dot{\xi}_{as} \\ \ddot{\xi}_{as} \\ \dot{\eta}_e \\ \ddot{\eta}_e \\ \dot{x}_L \end{Bmatrix} = \left[\begin{array}{cc|ccc} A_{\xi_{as}} & A_{\dot{\xi}_{as}} & 0 & 0 & 0 \\ A_{rr_0} & A_{rr_1} & A_{re_0} & A_{re_1} & A_{rL} \\ \hline 0 & 0 & 0 & I & 0 \\ A_{er_0} & A_{er_1} & A_{ee_0} & A_{ee_1} & A_{eL} \\ E_{L*} & E_{Lr} & 0 & E_{Le} & \frac{V}{L}R \end{array}\right] \begin{Bmatrix} \xi_{as} \\ \dot{\xi}_{as} \\ \eta_e \\ \dot{\eta}_e \\ x_L \end{Bmatrix} + \left[\begin{array}{ccc} 0 & 0 & 0 \\ B_{r_0} & B_{r_1} & B_{r_2} \\ \hline 0 & 0 & 0 \\ B_{e_0} & B_{e_1} & B_{e_2} \\ 0 & E_{L\delta} & 0 \end{array}\right] \begin{Bmatrix} \delta_u \\ \delta_{\dot{u}} \\ \delta_{\ddot{u}} \end{Bmatrix} + \left[\begin{array}{cc} 0 & 0 \\ B_{r_{w1}} & B_{r_{w2}} \\ \hline 0 & 0 \\ B_{e_{w1}} & B_{e_{w2}} \\ 0 & E_{rG} \end{array}\right] \begin{Bmatrix} w_G \\ \dot{w}_G \end{Bmatrix} \tag{34}$$

where $A_{\xi_{as}}$, $A_{\dot{\xi}_{as}}$ and E_{L*} are coupling matrices due to the similarity transformation executed. Now, the aeroelastic incremental loads, Δ_F and Δ_M, should be implemented in a way to allow a seamless integration between the nonlinear flight dynamics and the linear aeroelastic EoMs. In fact, this can be easily achieved in accordance with the partitions showed in Eq. (34) between rigid, elastic and aerodynamic lag dynamics. Hence, the aeroelastic incremental loads are computed similarly to the approximation given by Eq. (30),

$$\begin{bmatrix} \Delta_F \\ \Delta_M \end{bmatrix} = q_\infty \left\{ A_{0_{re}} \bar{\eta}_e + A_{1_{re}} \frac{L}{V} \dot{\eta}_e + A_{2_{re}} \frac{L^2}{V^2} \ddot{\eta}_e + D_{re} x_{Le} \right\} \tag{35}$$

Clearly to implement this algebraic equation, the generalized coordinate, $\bar{\eta}_e = \eta_e - \eta_{e_0}$, its rate, $\dot{\eta}_e$, and acceleration, $\ddot{\eta}_e$, vectors as well as the aerodynamic lag terms related with the elastic modes, x_{L_e}, must be estimated at each time iteration.

Note that the Minimum-State method is formulated to only use a single set of lag states, x_L, in Eq. (32). Therefore, the following augmented equation is devised to decouple the generalized coordinates' aero lag terms from the airframe states, ξ_{as} and $\dot{\xi}_{as}$, related ones,

$$\begin{Bmatrix} \dot{x}_{L_{as}} \\ \dot{x}_{L_e} \end{Bmatrix} = \begin{bmatrix} \frac{V}{L}R & 0 \\ 0 & \frac{V}{L}R \end{bmatrix} \begin{Bmatrix} x_{L_{as}} \\ x_{L_e} \end{Bmatrix} + \begin{bmatrix} E_{L*} & E_{Lr} & 0 & E_{Le} \\ 0 & 0 & 0 & E_{Le} \end{bmatrix} \begin{Bmatrix} \xi_{as} \\ \xi_{\dot{a}s} \\ \eta_e \\ \dot{\eta}_e \end{Bmatrix}$$

$$+ \begin{bmatrix} 0 & 0.5E_{L\delta} & 0 \\ 0 & 0.5E_{L\delta} & 0 \end{bmatrix} \begin{Bmatrix} \delta_u \\ \delta_{\dot{u}} \\ \delta_{\ddot{u}} \end{Bmatrix} + \begin{bmatrix} 0 & 0.5E_{rG} \\ 0 & 0.5E_{rG} \end{bmatrix} \begin{Bmatrix} w_G \\ \dot{w}_G \end{Bmatrix} \tag{36}$$

$$\begin{Bmatrix} z_{L_{as}} \\ z_{L_e} \end{Bmatrix} = \begin{bmatrix} D & D \end{bmatrix} \begin{Bmatrix} x_{L_{as}} \\ x_{L_e} \end{Bmatrix}$$

In this way, only elastic lag terms are considered to avoid any possible coupling with the rigid-body airframe related states (i.e. ξ_{as} and $\xi_{\dot{a}s}$). Now, the following differential equation is obtained combining the lower partition of Eq. (34) with the new devised Eq. (36).

$$\underbrace{\begin{Bmatrix} \dot{\eta}_e \\ \ddot{\eta}_e \\ \dot{x}_{L_{as}} \\ \dot{x}_{L_e} \end{Bmatrix}}_{\dot{x}_e} = \underbrace{\left[\begin{array}{cc|cc} 0 & I & 0 & 0 \\ A_{ee_0} & A_{ee_1} & A_{eL} & A_{eL} \\ \hline 0 & E_{Le} & \frac{V}{\bar{c}}R & 0 \\ 0 & E_{Le} & 0 & \frac{V}{\bar{c}}R \end{array}\right]}_{A} \underbrace{\begin{Bmatrix} \eta_e \\ \dot{\eta}_e \\ x_{L_{as}} \\ x_{L_e} \end{Bmatrix}}_{x_e} + \underbrace{\left[\begin{array}{ccc} 0 & 0 & 0 \\ B_{e_0} & B_{e_1} & B_{e_2} \\ \hline 0 & 0.5E_{L\delta} & 0 \\ 0 & 0.5E_{L\delta} & 0 \end{array}\right]}_{B_1} \underbrace{\begin{Bmatrix} \delta_u \\ \delta_{\dot{u}} \\ \delta_{\ddot{u}} \end{Bmatrix}}_{\delta_U}$$

$$+ \underbrace{\left[\begin{array}{cc} 0 & 0 \\ A_{er_0} & A_{er_1} \\ \hline E_{L*} & E_{Lr} \\ 0 & 0 \end{array}\right]}_{B_2} \underbrace{\begin{Bmatrix} \delta_{\xi_{as}} \\ \delta_{\dot{\xi}_{as}} \end{Bmatrix}}_{\delta_\xi} + \underbrace{\left[\begin{array}{cc} 0 & 0 \\ B_{ew1} & B_{ew2} \\ \hline 0 & 0.5E_{rG} \\ 0 & 0.5E_{rG} \end{array}\right]}_{B_3} \underbrace{\begin{Bmatrix} w_G \\ \dot{w}_G \end{Bmatrix}}_{w} \tag{37}$$

where $\delta_{\xi_{as}}$ and δ_u are defined as the incremental airframe states and inputs (perturbation from trim values),

$$\delta_{\xi_{as}} = \xi_{as} - \xi_{as|_0} \tag{38}$$

$$\delta_u = u - u_{|0} \tag{39}$$

$\xi_{as|_0}$ and $u_{|0}$ being the airframe state and input vectors computed at some specific trim condition *ab-initio* of the simulation run. Using a short notation form, Eq. (37) can be expressed as,

$$\dot{x}_e = A\, x_e + B_1\, \delta_U + B_2\, \delta_\xi + B_3\, w \tag{40}$$

The previous equation is used to estimate the elastic and lag states as a function of the incremental control input, ($\delta_U^T = [\ \delta_u^T,\ \delta_{\dot{u}}^T, \delta_{\ddot{u}}^T\ ]^T$), and incremental airframe states, ($\delta_{\xi}^T = [\ \delta_{\xi_{as}}^T,\ \delta_{\dot{\xi}_{as}}^T\ ]^T$) at each time iteration.

The quasi-static elastic deformation , η_{e0}, is computed by static residualization of the elastic modes, that is the $\dot{\eta}_e = \ddot{\eta}_e = x_{Le} = 0$ condition needs to be fulfilled. Therefore, the quasi-static elastic influence is estimated from Eq. (40) as:

$$\dot{x}_e = 0 \Longrightarrow x_e = -A^{-1}(B_1\, \delta_U + B_2\, \delta_x i + B_3\, w) \tag{41}$$

and from x_e the quasi-static elastic influence vector, η_{e0}, can be recovered.

In summary, the linear aeroelastic solver will be built based on:

1. Algebraic Eq. (35) to compute the incremental aeroelastic loads, Δ_F and Δ_M.
2. First-order differential Eq. (40) to compute the generalized coordinates related vectors $\bar{\eta}_e$, $\dot{\eta}_e$, and $\ddot{\eta}_e$, as well as the aerodynamic lag terms related with the elastic modes, x_{L_e}.
3. Algebraic Eq. (41) to estimate the quasi-static deformation vector η_{e_0} at that specific flight condition.

Fig. 7 illustrates the interconnection of the `F/A-18 AAW 6-DOF Dynamics` subsystem and the `Incremental Aeroelastic Solver`, `Control Surface Mixer` and `Control Command Transform` blocks.

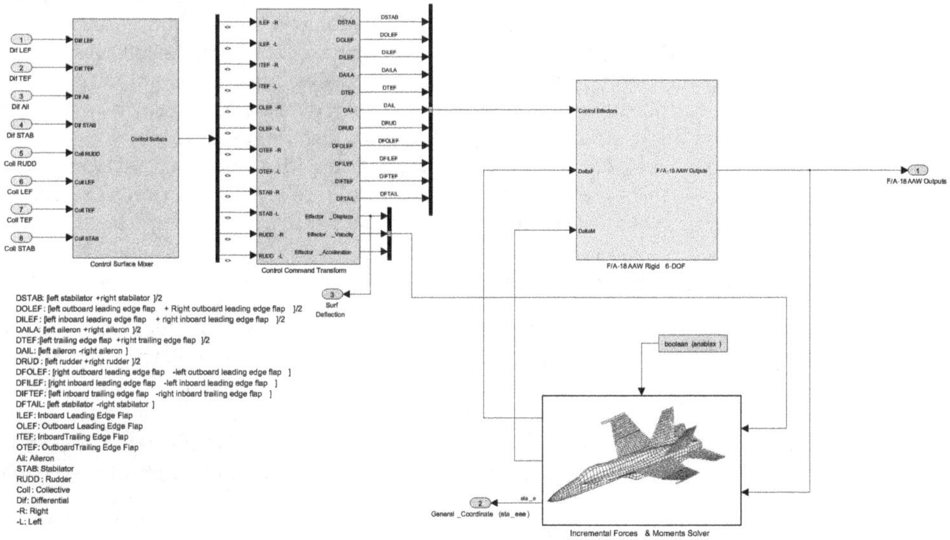

Fig. 7. Addition of the Incremental Aeroelastic Loads Solver to the Nonlinear Rigid-Body 6-DOF Subsystem.

6.2 Closed loop F/A-18 AAW linear model with gust excitation

In order to demonstrate the proposed feedforward filter design algorithm, a simplified closed loop F/A-18 AAW linear simulink model with gust excitation is developed/implemented for the evaluation purposes. This high-fidelity aeroelastic model was developed using the following elements:

- Six-degree-of-freedom solver using Euler angles subsystem.
- The AAW flight control system.
- Actuators and sensors.
- Aerodynamic Forces and Moments subsystem using the set of non-dimensional stability and control derivatives obtained through a set of AAW parameter identification flight tests.
- An incremental aeroelastic load solver including gust excitation generated by the ZAERO software system using rational function approximation techniques.

For continuous vertical gust perturbation, a low pass filter followed by a Dryden vertical velocity shaping filter is used to shape the power of the gust perturbation. The low pass filter is used to obtain the derivative of the gust perturbation. The low pass filter is given as $T_{LPF}(s) = \frac{a}{s+a}$ where $a = 200\pi\ rad/s$ is chosen in the remainder of the section.

The Dryden vertical velocity shaping filter is given as

$$T_g(s) = \sigma_{w_G} \frac{\sqrt{3}\tau_g^{-1/2}s + \tau_g^{-3/2}}{(s + 1/\tau_g)^2}$$

where $\tau_g = L_g/V$, and $L_g = 1750ft$, V is the aircraft body axis velocity; $\sigma_{w_G} = 100ft/s$.

For a more detailed development of the F/A-18 AAW simulink model with gust excitation, please refer to the NASA SBIR Phase I final report (Zeng & de Callafon, 2008). The implementation of the adaptive feedforward control algorithm to the linearized F/A-18 AAW simulink model is illustrated in Fig. 8. It should be noted that during the simulation study considered in this paper, the dynamics of airborne LIDAR turbulence sensor has not been considered. We assumed that a perfect gust perturbation can be measured by the airborne LIDAR turbulence sensor, i.e., the sensor dynamics has an ideal constant dynamics of 1. However, the practical effects of the airborne LIDAR turbulence sensor on the performance of the feedforward controller has to be addressed in the future study.

6.3 Implementation of the adaptive feedforward control

The construction of the feedforward controller can be separated into two steps, initialization and the recursive estimation of the filter. In the initialization step, a secondary path transfer function, $G(q)$, is estimated, which is done by performing an experiment using an external signal injected into the left and right trailing edge flaps as the excitation signal, and the error signal, $e(t)$, as the output signal. Since, $\hat{G}(q)$, is only used for filtering purposes, a high order model can be estimated to provide an accurate reconstruction of the filtered input, $\hat{u}_f(t)$, via

$$\hat{u}_f(t) = \hat{G}(q)w(t) \tag{42}$$

as described in Eq. (4).

To facilitate the use of the ORTFIR filter, a set of modal parameters need to be extracted to build the ORTFIR filter, using the frequency domain PolyMAX identification methodology in Section 4. With $\hat{u}_f(t)$ given in Eq. (42) and $d(t) = H(q)w(t)$ in place, the modal parameters can be easily estimated using the PolyMAX method. With the signal, $d(t)$, $\hat{u}_f(t)$ and the basis function, $B_i(q)$, a recursive minimization of the feedforward filter is done via the recursive least squares minimization technique described in Section 5.

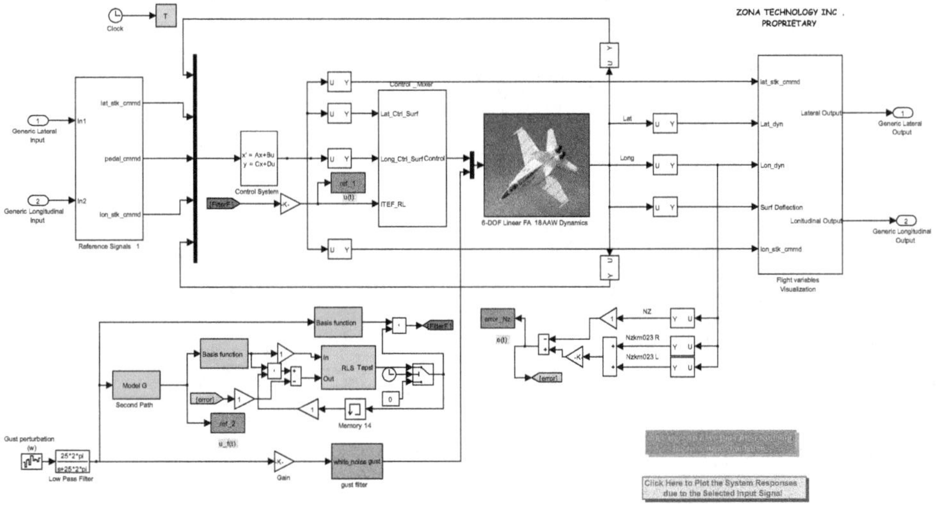

Fig. 8. Closed Loop F/A-18 AAW Linear Simulink Model.

The error signal, $e(t)$, can be selected as the vertical accelerometer reading at the aircraft left/right wing folder positions, i.e., Nz_{km023R} or Nz_{km023L}. An alternative choice could be

$$e(t) = \left[\frac{Nz_{km023R} + Nz_{km023L}}{2} - Nz_{cg} \right] \tag{43}$$

In this paper, Eq. (43) is served as a feedback signal for feedforward filter design purpose. The advantage of choosing Eq. (43) is that the rigid body dynamics can be partly removed, and the vertical wing bending is still observed.

Upon initialization of the feedforward controller, a 20th order ORTFIR model, $\hat{G}(q)$, was estimated in order to create the filtered signal, $\hat{u}_f(t)$. The amplitude bode plot of the estimated, $\hat{G}(q)$, is shown in Fig. 9.

The modes used to build the orthonormal basis, $B_i(q)$ are extracted from the stabilization diagram in Fig. 10. From Fig. 10, five elastic modes can be extracted, and they are shown in Table 4.

	PolyMAX Identification	
Mode Number	Frequency [Hz]	Damping ζ [%]
1	5.9246	4.5311
2	10.083	4.182
3	13.602	10.072
4	18.377	2.7409
5	21.569	2.5183

Table 4. Estimated Modes of Feedforward Filter Using FFT-PolyMAX Method.

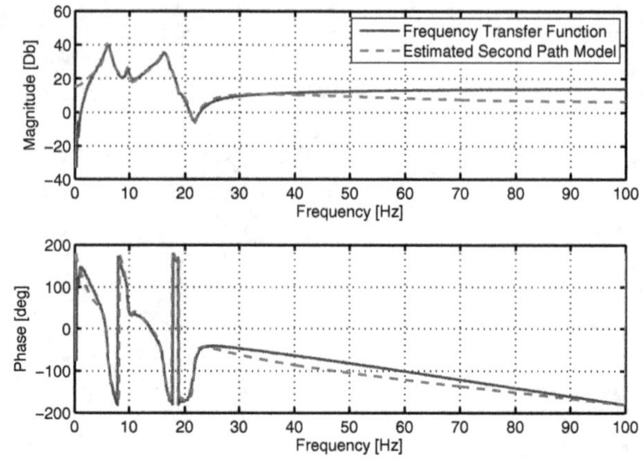

Fig. 9. Bode Plot of the Estimated 20th Order Secondary Path Model, $\hat{G}(q)$.

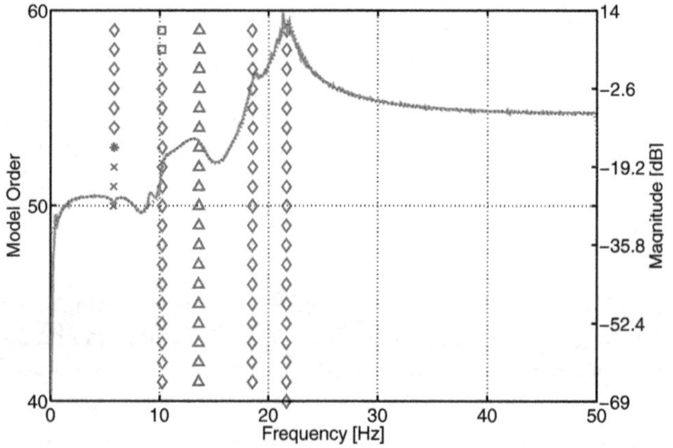

Fig. 10. Stabilization Diagram.

For implementation purposes, only $L = 2$ parameters in the ORTFIR filter are estimated. With a 10th order basis, $B_i(q)$, this amounts to 20th order ORTFIR filter. To evaluate the performance of the proposed ORTFIR filter for feedforward compensation, a 20th order Finite Impulse Response Filter is also designed to reduce the vertical wing vibration.

For a clear performance comparison between FIR filter and ORTFIR filter, the frequency response of the Nz_{km023R}, and Nz_{km023L} using FIR filter and ORTFIR filter are plotted in Fig. 11 and Fig. 12, respectively. The solid line in Fig. 11 (a) is the auto spectrum of the accelerometer measurement Nz_{km023R} without feedforward controller integrated in the system; the dashed line in Fig. 11(a) indicates the auto spectrum of the accelerometer measurement Nz_{km023R} with the adaptive feedforward controller using FIR filter added in the system; the dotted line in Fig. 11 shows the auto spectrum of the accelerometer measurement Nz_{km023R} with the adaptive feedforward controller using ORTFIR filter added in the system. Fig. 11 (b) is the

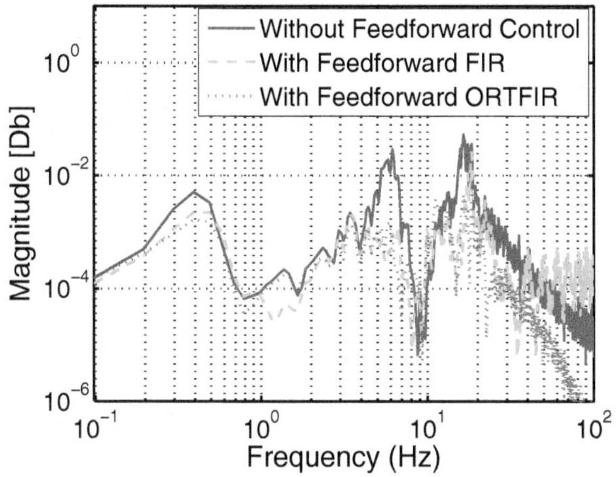

(a) Frequency Spectrum Plot of the Nz_{km023R} .

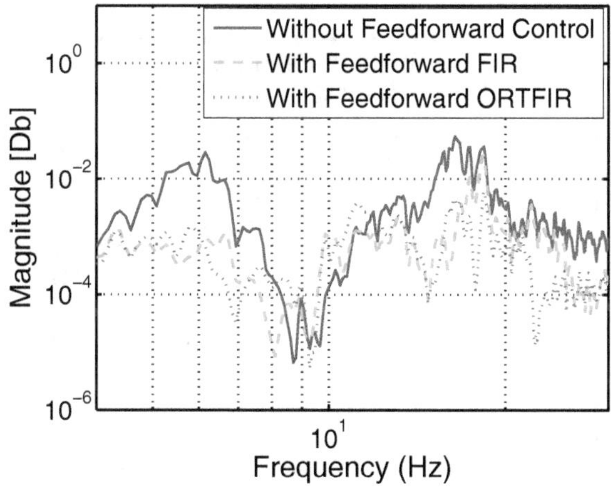

(b) Zoomed Frequency Spectrum Plot of the Nz_{km023R}.

Fig. 11. Spectral Content Estimates of the Nz_{km023R} Without Control (Solid), With Control Using 20th Order FIR Filter (Dashed), and Using 20th Order ORTFIR Filter (Dotted).

zoomed-in plot of Fig. 11 (a) in the frequency range of [4 30] Hz. It is clearly seen that with the ORTFIR filter, a better magnitude reduction of auto spectrum of Nz_{km023R} can be obtained in most of the frequency range compared to the FIR filter. Similar performance could also be observed in regards to Nz_{km023L}, and which is shown in Fig. 12 (a) and (b).

The corresponding time responses are illustrated in Fig. 13 and Fig. 14. Fig. 13 (b) and Fig. 14 (b) are the zoomed-in plots of Fig. 13 (a) and Fig. 14 (a), respectively. These time responses clearly demonstrate that with the adaptive feedforward controller using ORTFIR filter, a better structural vibration reduction can be obtained. From these figures, it is clearly demonstrated

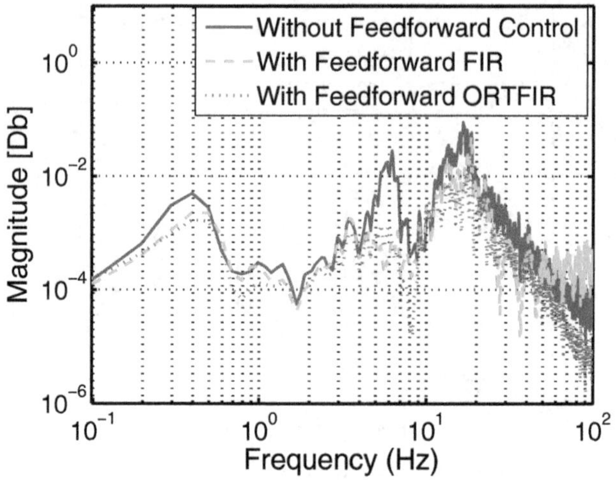

(a) Frequency Spectrum Plot of the Nz_{km023L} .

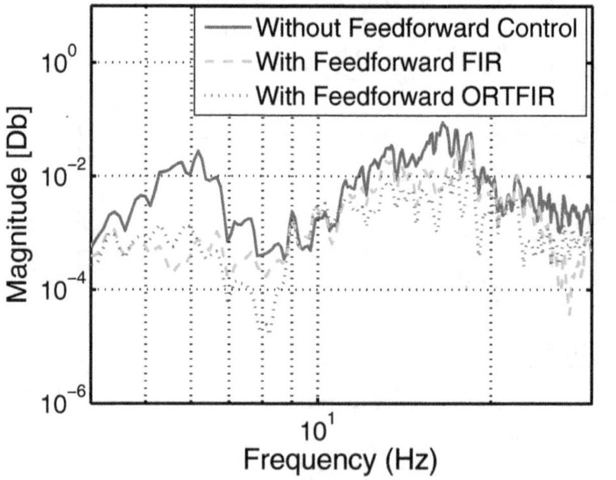

(b) Zoomed Frequency Spectrum Plot of the Nz_{km023L}.

Fig. 12. Spectral Content Estimates of the Nz_{km023L} Without Control (Solid), With Control Using 20th Order FIR Filter (Dashed), and Using 20th Order ORTFIR Filter (Dotted).

that both FIR filter and ORTFIR filter are efficient to reduce the normal acceleration at the left wing folder position and right wing folder position. With the use of the both the ORTFIR filter and FIR filter, the spectral content of the Nz_{km023R} and Nz_{km023L} have been reduced significantly in the frequency range from 2Hz to 20Hz. However, with the use of the ORTFIR filter, more efficient vibration reduction performances are expected compared to the FIR filter.

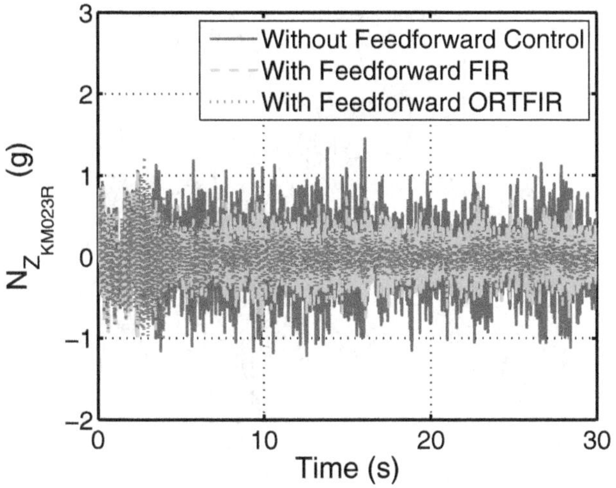

(a) Time Domain Response Plot of the Nz_{km023L}.

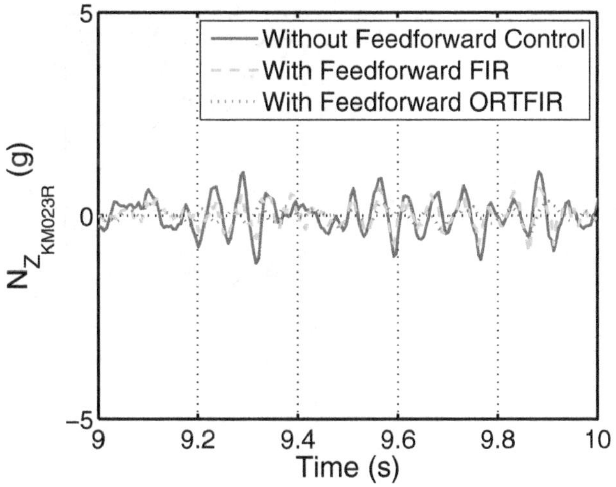

(b) Zoomed Time Domain Response Plot of the Nz_{km023L}.

Fig. 13. Time domain Response of the Nz_{km023R} Without Control (Solid), With Control Using 20th Order FIR Filter (Dashed), and Using 20th Order ORTFIR Filter (Dotted).

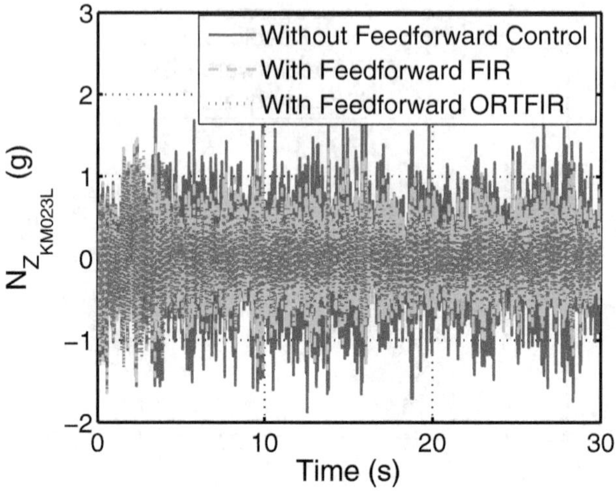

(a) Time Domain Response Plot of the Nz_{km023L} .

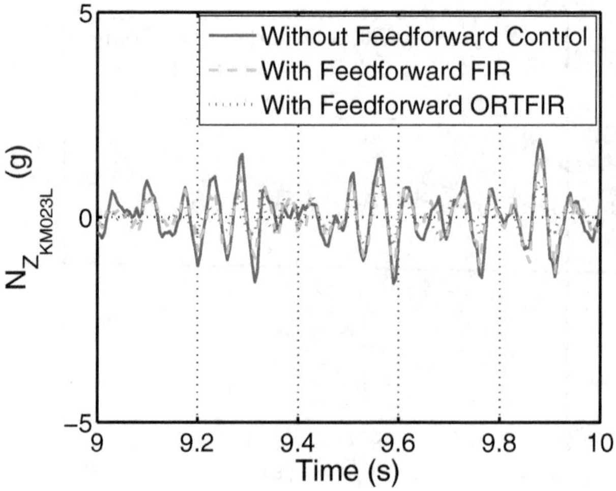

(b) Zoomed Time Domain Response Plot of the Nz_{km023L}.

Fig. 14. Time Domain Response of the Nz_{km023L} Without Control (Solid) and Using 20th Order FIR Filter (Dashed), and With Control Using 20th Order ORTFIR Filter (Dotted).

7. Conclusions

In this chapter, an adaptive feedforward control methodology has been proposed for the active control of gust loads alleviation using an ORTFIR filter. The ORTFIR filter has the same linear parameter structure as a taped delay FIR filter that is favorable for (recursive) estimation. The advantage of using the ORTFIR filter is that it allows the inclusion of prior knowledge of the flexible mode information of the aircraft dynamics in the parametrization of the filter for better accuracy of the feedforward filter.

In addition, by combining the flight dynamics model for rigid body dynamics and an aeroelastic solver for aeroelastic incremental loads to accurately mimic in-flight recorded dynamic behavior of the air vehicle, a unified integration framework that blends flight dynamics and aeroelastic model is developed to facilitate the pre-flight simulation.

The proposed methodology in this chapter is implemented on an F/A-18 AAW aeroelastic model developed with the unified integration framework. The feedforward filter is updated via the recursive least square technique with the variable forgetting factor at each time step. Compared with a traditional FIR filter and evaluated on the basis of the simulation data from the F/A-18 AAW aeroelastic model, it demonstrates that applying the adaptive feedforward controller using the ORTFIR filter yields a better performance of the gust loads alleviation of the aircraft.

8. References

Andrighettoni, M. & Mantegazza, P. (1998). Multi-Input/Multi-output adaptive active flutter suppression for a wing model, *Journal of Aircraft* 35(3): 462–469.

Baldelli, D. H., Chen, P. C. & Panza, J. (2006). Unified aeroelastic and flight dynamics formulation via rational function approximations, *Journal of Aircraft* 43(3): 763–772.

Baldelli, D. H. & Zeng, J. (2007). Unified nonlinear flight dynamics aeroelastic solver tool, *Technical Report SBIR Phase I NNL07AA85P*, ZONA Technology Inc.

Baldelli, D. H., Zeng, J., Lind, R. & Harris, C. (2009). Flutter-prediction tool for flight-test-based aeroelastic parameter-varying models, *Journal of Guidance, Control, and Dynamics* 32(1): 158–171.

Barker, J. M., Balas, G. J. & Blue, P. A. (1999). Active flutter suppression via gain-scheduled linear fractional control, *Proceedings of the American Control Conference*, San Diego, California, pp. 4014–4018.

Cauberghe, B., Guillaume, P., Verboven, P., Parloo, E. & Vanlanduit, S. (2004). A poly-reference implementation of the maximum likelihood complex frequency-domain estimator and some industrial applications, *Proceedings of the 22nd International Modal Analysis Conference*, Dearborn, US.

Cauberghe, B., Guillaume, P., Verboven, P., Vanlanduit, S. & Parloo, E. (2005). On the influence of the parameter constraint on the stability of the poles and the discrimination capabilities of the stabilization diagrams, *Mechanical Systems and Signal Processing* 19(5): 989–1014.

Eversman, W. & Roy, I. D. (1996). Active flutter suppresion using MIMO adaptive LMS control, *37th AIAA / ASME / ASCE / AHS / ASC Structures, Structural Dynamics, and Material Conference and Exhibit*, Salt Lake, UT. AIAA-1996-1345.

Garrard, W. L. & Liebst, B. S. (1985). Active flutter suppression using eigenspace and linear quadratic design techniques, *Journal of Guidance, Control, and Dynamics* 8(3): 304–311.

Haykin, S. (2002). *Adaptive Filter Theory*, Prentice Hall, Englewood Cliffs, NJ.

Heuberger, P. S. C., Van Den Hof, P. M. J. & Bosgra, O. H. (1995). A generalized orthonormal basis for linear dynamical systems, *IEEE Transactions on Automatic Control* 40(3): 451–465.

Karpel, M. (1992). Size reduction techniques for the determination of efficient aeroservoelastic models, *Control and Dynamic Systems* 54: 263–295.

Leibst, B. S., Garrard, W. L. & Farm, J. A. (1988). Design of a multivariable fluter suppression/gust load alleviation system, *Journal of Guidance, Control, and Dynamics* 11(3): 220–229.

Ljung, L. (1999). *System Identification: Theory for the User*, Prentice-Hall, Englewood Cliffs, NJ.

Mahesh, J. K., Stone, C. R., Garrard, W. L. & Dunn, H. J. (1981). Control law synthesis for flutter suppression using linear quadratic control theory, *Journal of Guidance, Control, and Dynamics* 4(4): 415–422.

Morelli, E. A. (1995). Global nonlinear aerodynamic modeling using multivariate orthogonal functions, *Journal of Aircraft* 32(2): 270–277.

Morelli, E. A. & Klein, V. (1997). Accuracy of aerodynamic model parameters estimated from flight test data, *Journal Guidance, Control, and Dynamics* 20(1): 74–80.

Newsom, J. R. (1979). Control law synthesis for active flutter suppresion using optimal control theory, *Journal of Guidance, Control, and Dynamics* 2(5): 388–394.

Ninness, B. & Gustafsson, F. (1997). A unifying construction of orthonormal bases for system identification, *IEEE Transactions on Automatic Control* 42(4): 515–521.

Park, D. J. (1991). Fast tracking RLS algorithm using novel variable forgetting factor with unity ZONE, *Electronic Letter* 27: 2150–2151.

Schmidt, D. & Chen, T. (1986). Frequency domain synthesis of a robust flutter suppression control law, *Journal of Guidance, Control, and Dynamics* 9(3): 346–351.

Schmitt, N. P. Amd Rehm, W., Pistner, T., Zeller, P., Diehl, H. & Navé, P. (2007). The AWAITOR airborne LIDAR turbulence sensor, *Aerospace Science and Technology* 11: 546–552.

Schmitt, N., Rehm, W., Pistner, T., Diehl, H. Navé, P., Jenaro-Rabadan, G., Mirand, P. & Reymond, M. (2007). Flight test of the AWAITOR airborne LIDAR turbulence sensor, *14th Coherent Laser Radar Conference*, Colorado.

Wildschek, A., Maier, R., Hoffmann, F., Jeanneau, M. & Baier, H. (2006). Active wing load alleviation with an adaptive feedforward control algorithm, *AIAA Guidance, Navigation, and Control Conference and Exhibit*, Keyston, Colorado. AIAA-2006-6054.

Woods-Vedeler, J. A., Pototzky, A. & Hoadley, S. T. (1995). Rolling maneuver load alleviation using active controls, *Journal of Aircraft* 32(1): 68–76.

Zeng, J. & de Callafon, R. A. (2006). Model matching and filter design using orthonormal basis functions, *45th IEEE Conference on Decision and Control*, San Diego, pp. 5347–5352.

Zeng, J. & de Callafon, R. A. (2008). Adaptive Feedforward/Feedback control framework, *Technical Report SBIR Phase I NNX08CB12P*, ZONA Technology Inc.

6

Effects of Automatic Flight Control System on Chinook Underslung Load Failures

Marilena D. Pavel
Faculty of Aerospace Engineering, Delft University of Technology
The Netherlands

1. Introduction

One of the major helicopter attributes is its ability to transport cargo externally in the form of external slung loads (see Fig. 1). Commercial and military operators accept the fact that using a helicopter for external load transportation is usually expensive in terms of both money and time. However, helicopters still have the significant advantage of accessing unreachable sites. Operations of helicopters with external loads impose limitations to the use of the helicopter, as for example: helicopter maximum forward speed is usually severely reduced because of the danger on dynamic instabilities of the load; due to external load the aerodynamic drag can become excessive, resulting in power and control limitations on the helicopter. The problem addressed in this chapter concerns the behaviour of a helicopter following the premature breakdown of one of its cables sustaining the slung load. As a specific example, the Chinook helicopter CH-47B with an external load will be considered. The CH-47 (Chinook) is a twin-engine tandem rotor helicopter (see Fig. 1) designed for all-weather, medium-sized transport type operations. The three bladed rotors are driven in opposite directions (front rotor rotates anticlockwise, rear rotor rotates clockwise) through interconnecting shafts which enable both rotors to be driven by either engine. The rotor heads are fully articulated, with pitch, flapping, and lead-lag hinges.

Fig. 1. Chinook helicopter transporting single and multiple loads (Courtesy of the Royal Netherlands Air Force)

The goal of the chapter will be to implement the advanced automatic flight control system AFCS of the Chinook CH-47 (analogue in Chinook D-version and digital in the latest version F) into a generic Chinook model developed at Delft University of Technology (Van der Kamp et. Al (2005), Reijm, Pavel&Bart (2006), Pavel(2007), Pavel(2010)) and investigate the effects that AFCS may have on the recovery prospects of the Chinook helicopter after a failure scenario of its load. In other words, this chapter proposes to analyse how the advanced AFCS, implemented in general in order to improve the handling qualities characteristics, may improve/degrade the CH-47 behaviour during emergency situations such as failure scenarios of its suspended load(s). Searching in the specialist literature for research on this particular area of load failures and the AFCS effects on assisting/hindering the pilot during load failure recovering revealed that the subject has not been really investigated. A few publications have been identified dealing with load failure scenarios - they address mainly a rather different problem, i.e. the situation in which a load is moving within an aircraft before being dropped - and none considers the effects that AFCS may have on helicopter behaviour during such emergency situations. Some relevant publications for understanding the problems related to the dynamics of a helicopter with a slung load are given below.

(Lucasen and Sterk, 1965) developed a first theoretical study of the dynamics of a single rotor helicopter carrying a slung load in hovering flight using a 3-dof longitudinal helicopter model including translation and attitude of the helicopter and load displacement w.r.t. the helicopter. They demonstrated that the phugoid stability depended on the cable length, decreasing with increasing cable length. (Dukes, 1973a) studied the stability characteristics of a single rotor helicopter with slung load near hover showing that with large pitch damping, the translational motion is only weakly coupled to the attitude and load motions. In a second study (Dukes, 1973b) concentrated on the stability of the load during different manoeuvres (acceleration-deceleration, changing the hover location and arresting a pendulous load motion) identifying some fundamental features of load control which can be utilized as basis for an open loop control strategy. (Feaster, Poli & Kirchhoff, 1977) studied the stability in forward flight of a single-rotor helicopter carrying a container, showing that long cables, high speeds and low weights increased the stability of the loads. Their results were contradicted by (Cliff&Bailey,1975) who obtained that longer cables were destabilizing, this probably due to different load aerodynamics considered. Both last papers underlined the importance of the load aerodynamics in studying the stability of helicopter-slung load systems, where dynamic instabilities can be triggered by unsteady load aerodynamics. The most dominant form of load aerodynamic instability is a yawing divergence that couples with the load lateral modes. (Gabel, 1968) studied the slung load "vertical bounce" phenomenon, i.e. a resonant condition inherent to elastic sling load systems that occurs when the natural frequency of the sling load is close to one of the rotor frequencies. This mode can be exacerbated by the pilot's inertial reaction to vertical accelerations, leading to a vertical force being transmitted to the collective stick. To account for vertical bounce phenomenon one has to consider elastic suspensions. The author proposed several solutions to the pilot induced vertical bounce problem, the most promising being a low gain boost system which was meant to produce very low output for small pilot inputs and normal output for normal pilot inputs. For forwards speeds above 40 knots, the author showed that the effects of rotor downwash on the load aerodynamics are not important. (Nagabhushan, 1985) developed a non-linear body-flap model of a single rotor-helicopter and studied the stability characteristics of the helicopter-slung load system. He found that operating with a long sling cable damps out the sling-load pendulum mode lateral oscillation and the helicopter longitudinal phugoid; however, the associated sling-load pitch oscillation and the helicopter Dutch roll mode become unstable. Suspending the sling load from a point ahead of the helicopter centre of gravity was found to stabilize the helicopter lateral oscillation; suspending it from a point aft of the centre of gravity caused instability. (Sheldon, 1978) concluded from experiments that a large number of instabilities are initiated by load yawing motions. A relatively easy way of obtaining an increased yaw resistance is to use a multipoint suspension system rather than a single point suspension. His experiments have shown that an inverted 'V' suspension system provides significant yaw resistance, while also minimizing the load trail angle. (Prabhakar, 1978) developed a non-linear single rotor helicopter-slung load model in multipoint suspension showing that the number, spacing and placement of the suspension points and the topology of the suspension cables are all important. The load is capable of changing almost completely the helicopter rigid body modes, generally destabilizing. Also, there exists significant coupling between the load and the helicopter lateral modes. The helicopter longitudinal modes on the other hand are affected little by the presence of the load. Dynamic coupling between the Dutch roll and a lateral load mode has been shown to decrease the Dutch role damping. A more recent study of (Cicolani, Kanning& Synnestvedt, 1995) developed a comprehensive approach for slung-load modelling where generic simulation models for a tandem helicopter capable of carrying load in single or multiple points with one, two or more helicopters were investigated. (Fusato, Guglieri & Celi, 2001) developed a body-flap-lag model to study the flight dynamics of a single-rotor helicopter carrying a single-point suspended load, showing that the load affects trim primarily through the overall increase in the weight of the aircraft; the influence of cable length was negligible. (Stuckey, 2002), using the equations of motions developed by (Cicolani, Kanning& Synnestvedt, 1995) and an open-loop control for pilot modelling, developed a piloted simulation model for a tandem helicopter capable of carrying a mixed density slung loads. (Tyson.1999) developed a slung-load simulation model composed on the GenHel UH-60 model and validated against flight test data. (Chen, 1998) built a non-linear simulation model for one rotor helicopter-slung system and investigated the sudden load movement causing the cables to slacken or collapse. (Kendrick&Walker, 2006), using the motion simulator at The University of Liverpool, investigated the stability and handling qualities of a single-rotor helicopter carrying a slung load showing that at hover and low speed the pilot found easier to fly the helicopter with external load due to the increased damping in pitch provided by the load. (Bisgaard, 2006) developed an intuitive and easy-to-use way of modelling and simulating different slung load suspension types by using a redundant coordinate formulation based on Gauss' principle of Least Constraint using the Udwadia-Kalaba equation.

(Van der Kamp et. Al, 2005) (Reijm, Pavel&Bart, 2006) and (Pavel, 2010) investigated whether the three-strop suspension system - usually used to transport large loads (see right-hand side in Fig. 1) can be safely replaced by a two-strop suspension. A three-strop suspension system is actually a two-strop suspension backed up by a third point, the so-called 'redundant HUSLE'. The redundant HUSLE is a redundant set of slings which comes into action if one of the normal strops fails. However, such a system is expensive in terms of both money and time. Therefore, questioning how reliable the two-strop suspension was when compared to the three-point suspension; the above references concluded that although, in general, flying with the redundant HUSLE resulted in less violent helicopter reactions after load failure, redundant HUSLE did not necessarily mean safer. More specifically, they demonstrated that loads up to only two tonnes could be safely suspended only in two-points without the need of using redundant HUSLE.

One of the many features of the CH-47 is its automatic flight control system AFCS. The goal of this chapter is to implement an advanced flight control system (AFCS) replicating the longitudinal axis AFCS of a Chinook CH-47D and use this system switched on to determine how the AFCS influences the recovery prospects of the Chinook helicopter after a front suspension failure. The chapter is structured as follows:

- The first section describes the AFCS characteristics;
- The second section develops the control laws for a longitudinal AFCS with pitch attitude hold and airspeed hold;
- The third section simulates an example of a failure scenario flown and determines the AFCS effects for recovering;
- The fourth section defines safety envelopes as boundaries of the maximum helicopter forward velocity achievable when recovery is possible after a load failure as a function of the load mass that can be safely transported;
- Finally, general conclusions and potential further extension of this work are discussed.

1.1 Tandem helicopter control

The pilot controls of a tandem helicopter are essentially the same as those for normal helicopters. A pilot has a stick to control longitudinal and lateral motion, a collective stick for thrust and pedals to control the yaw motion. However, the connections between controls and rotor hub angles are different. Collective input moves both swash plates equally up and down. Whereas a normal helicopter longitudinal control input tilts the swash plate longitudinally, this input results in differential collective in a tandem helicopter. Lateral motion is achieved by tilting both front and rear rotor in the same direction. The tail rotor that is used for yaw control, is missing. This function is replaced by lateral cyclic in opposite direction for front and rear rotor disc. It should be noted that pitch attitude is not only controlled by differential collective of the front and rear rotor heads. In the past many tests have been conducted in order to achieve acceptable pitch control. The first Piasecki tandem helicopter and the Bristol 173 (see Fig. 2) had longitudinal cyclic, both were either uncontrollable or could only fly backwards (Prouty, 2001). Therefore a later Piasecki model, the HRP-1 used differential collective for longitudinal control.

Fig. 2. Bristol 173 with cyclic pitch longitudinal control, Piasecki HRP-1 with differential pitch, www.aviastar.org

This type of helicopter control implies that with increased velocity the thrust of both rotors has to be tilted forward to compensate for the higher helicopter body drag. The lack of longitudinal cyclic in the HRP-1 forced the pilot to apply initially more differential thrust. This introduced an increase in nose down pitch angle. As a result of this forward helicopter pitch angle, the drag of the helicopter increases dramatically (see Fig. 3 from (Prouty, 2001)) To compensate for the large pitch angle, the longitudinal stick of the Chinook CH47-D controls differential thrust. Longitudinal cyclic is controlled separately by airspeed, keeping the helicopter body in acceptable pitch angles, thus minimizing helicopter body drag.

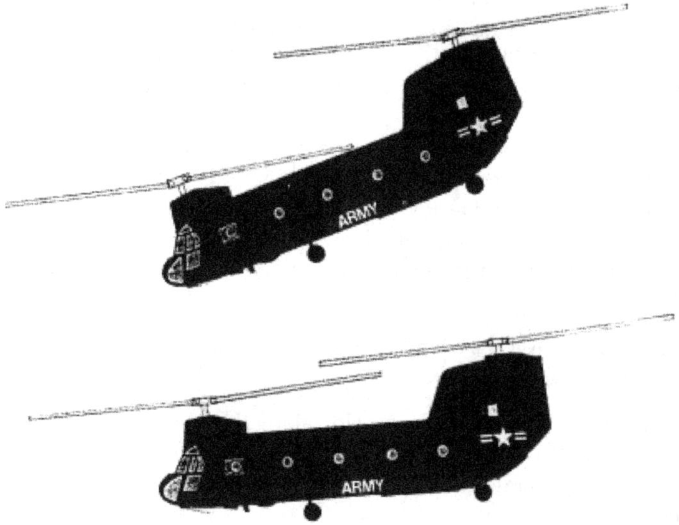

Fig. 3. Tandem helicopter with only differential pitch (top) and with differential collective and longitudinal cyclic pitch (bottom), (Prouty, 2001)

2. Description of the Chinook Automatic Flight Control System

An Advanced or Automatic Flight Control System (AFCS) employs the aerodynamic control surfaces to manipulate the forces and moments of an aircraft to control its velocity, attitude and position. The AFCS can do this with or without assistance from the pilot. The performance of the AFCS is largely governed by its flight control laws that translate the input of various sensors to control surface output. Just like the complete system, the design of these control laws, especially for military aircraft, is determined by the requirement to provide good handling qualities, over a wide range of operating conditions (including cargo transport), with a low pilot workload, while being easy to expand or modify.

For the Chinook helicopter, pilot control is done by varying the pitch of the blades either cyclically or collectively. This is done with the thrust control lever (collective), a cyclic control stick and the directional pedals, which are all coupled between the pilot and the co-pilot's position. The stick and pedal movements are transferred with a system of bell-cranks, push-pull tubes and actuators to a control mixing closet, located in the small hallway connecting the cabin to the cockpit. In this mixing unit, the pilot control inputs are combined with the signals from the AFCS computers and then mixed to result in the required lateral cyclic and collective pitch of the two rotors.

Earlier models of the CH-47 only had a Stability Augmentation System (SAS) installed to assist the pilot in attitude stabilization. With the D-version, the AFCS was added, which brought a number of modifications and additions to the SAS. Still, the main objective remains attitude stabilization, including rate damping, of the helicopter about all three axes. However, it is extended with a number of features, that for example maintain the desired value of certain flight parameters such as airspeed, pitch attitude and bank angle. The CH-47F is to be the first of the Chinook family equipped with a digital AFCS (DAFCS), granting it Level I handling qualities and meeting the stringent ADS-33 requirements for operation in degraded visual environments. A few of its impressive features include position adjustment with 30 cm increments, automatic hover capture when the cyclic stick is released below 8 kts ground speed and an altitude hold mode that eliminates drift (Einthoven et. Al., 2006). Since the flight control actuation on the F-version remains unchanged from the D-version, the control system still contains mechanically linked actuators. Hence it does not qualify for the term fly-by-wire: this notion originated from the replacement of mechanical linkages by electric signalling.

The main features of the AFCS system as implemented in the Chinook version D consist of (Boeing Helicopters, 2004):

- Pitch attitude and long term airspeed hold, taking the position of the longitudinal cyclic stick as a reference;
- Long term bank angle and heading hold in level flight, bank angle hold in turning flight (performing coupled turns);
- A stable positive longitudinal stick gradient throughout the entire flight envelope;
- Fine adjustment of bank angle and airspeed trim (Vernier beep trim);
- Altitude hold by means of barometer or radar signals (this mode is valid in the CH 47D of the US Army; the Royal Netherlands Air Force replaced this mode by Flight Director);
- Improved manoeuvrability with the use of control position transducers for all cockpit controls;
- Electronic detent switching on lateral stick and pedals based on signals supplied by the AFCS;
- Longitudinal cyclic trim scheduling and automatic LCT positioning to ground mode when the aft wheels touch the ground;
- Use of the HSI bug error (error between actual and desired heading, indicated by a small token on the HSI instrument) for heading select (only in the CH 47D of the US Army; the Royal Netherlands Air Force replaced this mode by Flight Director);

Fig. 4 presents the cockpit interior of a CH-47D. Looking at this figure one can identify: 1) the AFCS control panel; 2) co-pilot pitch-roll control (cyclic stick); 3) pilot thrust control; 4) AFCS computer; 5) co-pilot multifunction display MMFD; 6) instrument panel; 7) centre console; 8) pilot pitch-roll control (cyclic stick); 9) multifunction display (AVMFD).

Fig. 4. Cockpit interior of a CH-47D, (Boeing Helicopters, 2004)

Regarding the AFCS of a CH-47D, this consists of the following parts:

- A cockpit control panel
- Two AFCS computers
- Three integrated lower control actuators (ILCA's)
- Two differentials airspeed hold actuators (DASH)
- Two Longitudinal Cyclic Trim actuators (LCT)
- Attached to the cockpit controls are 2 magnetic brakes for yaw and roll, a longitudinal cockpit control driver actuator (CCDA) and a collective CCDA
- Three control position transducers (CPT's)

The AFCS cockpit control panel (number 1 in Fig. 4) contains 7 switches to control the many AFCS functions (see Fig. 5). On top, the flight director coupling switches that when pressed engage the coupling between the AFCS and the Flight Director. Only one flight director may be coupled at a time. The left lower side contains the switches that control the cyclic trim actuators. Normally the longitudinal cyclic trim (LCT) actuators are automatically operated by the AFCS computers, with computer 1 controlling the forward LCT and computer 2 governing the aft LCT actuator according to the LCT trim schedule. When the switch is flipped to manual setting, the pilot is able to independently direct the extension or retraction of both actuators. The AFCS system select switch can be found on the lower right side of the display. Usually both AFCS systems will be selected. When one AFCS system is switched off, the Flight Director coupling and LCT functions continue to operate. The latter even keeps working with neither AFCS in operation.

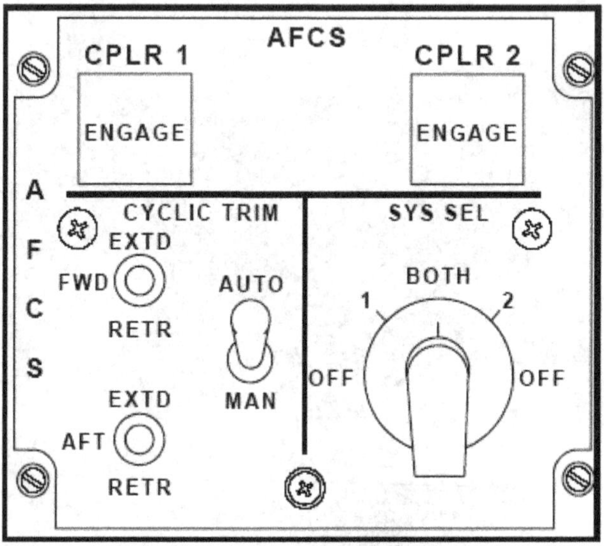

Fig. 5. AFCS control panel scheme, (Boeing Helicopters, 2004)

Two AFCS computers are located in the avionics compartment. Since the AFCS is a redundant system, in normal operation, each computer controls half of the input to the flight controls. This is described as "operation at half gain and half authority". Failure of one of the computers results in the other computer taking over at ¾ gain but like regular operation, it has only up to half of the maximum travel of the working system available. In that case the remaining system is said to function at ¾ gain and half authority (Anon., 2002). When operational, the computers receive flight data from sensors and convert this into command signals that are fed to the actuators. Each unit directs its signals to the actuators according to a different path. Each actuator provides a position feedback signal to the related AFCS computer.

The Integrated Lower Control Actuators (ILCA) (see Fig. 6(a)) span 3 channels: pitch, roll, yaw. They separate the pilot control forces from the forces required to move the swashplates, which are generated by the upper controls. The ILCA's consist of two parts: a lower boost actuator and a dual extensible link. The hydromechanical lower boost actuators assist the pilot in controlling the helicopter. They are mechanically linked to the cockpit controls and will extend or retract in response to pilot inputs. The hydroelectrical dual extensible links incorporate an upper link, controlled by AFCS computer 1, and a lower link, driven by computer 2. This means that for full travel (full authority) both computers have to supply input. The actuating cylinders move solely based on AFCS commands without any corresponding cockpit controller motion. In the case none of the AFCS computers are working, the extensible links act as rigid links. The thrust input is enhanced with a lower boost actuator without extensible links, thereby assisting the pilot in moving the thrust control without providing a hydroelectrical input for the AFCS.

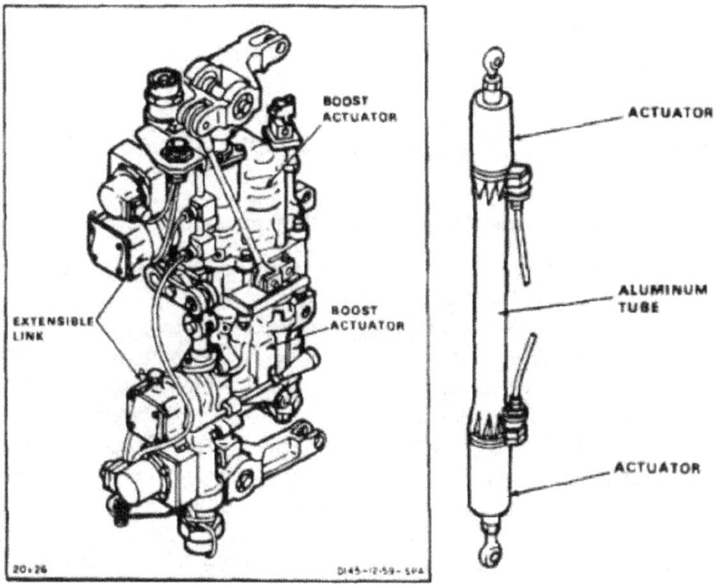

(a) Integrated Lower Control Actuator (b) Differential Airspeed Hold Actuator

Fig. 6. Chinook actuators connected in series to cockpit controls (Anon, 2002)

The differential airspeed hold (DASH) actuator (see Fig. 6(b)) is installed between the cyclic stick and the pitch ILCA. The actuator is in fact a combination of two electro-mechanical linear actuators mounted end to-end inside a tube. Like an ILCA, the upper half is controlled by AFCS No. 1, the lower half by No. 2, so again both AFCS computers are required for full authority, which is equal to 50% of the longitudinal control range, taking about 5 seconds to complete full travel. The DASH extends and retracts with the purpose of long term stabilization of the pitch axis, arranging a positive stick gradient and maintaining airspeed about a fixed stick position. Its performance depends on the mode of operation, of which there are three:

- A normal mode in which it stabilizes 1) the airspeed when flying faster than 40 kts and 2) the pitch attitude at speeds below 40 kts.
- A Differential Collective Pitch Trim (DCPT) mode. This is turned on in special conditions such as high cyclic stick rate, large pitch attitude or on the ground. This is to prevent an exaggerated response from the actuator that would occur in these instances with the DASH in normal mode.
- The transition mode. This mode is entered when the DCPT conditions do not apply anymore and the AFCS is engaged. The system switches to a low rate driver (20% of the normal rate), forcing the DASH actuator to return to its normal position corresponding to the airspeed. This avoids a step-like control input when the DASH resumes normal operation.

The two longitudinal cyclic trim (LCT) actuators installed under the swashplate have as primary task to minimize fuselage drag by reducing the nose down attitude as airspeed and altitude build up. To manage this, the AFCS transmits signals to the LCT actuators which in turn increase the longitudinal cyclic pitch angle of the fore and aft rotor. This way blade flapping is also reduced, lowering stresses on the rotor shafts. The pilot can also manually select the LCT actuator positions.

3. Modelling the Chinook tandem helicopter and its Automatic Flight Control System

3.1 Modelling the Chinook tandem helicopter with external load

A general non-linear six degree-of-freedom (6-dof) rigid body model for a generic tandem helicopter was first developed for piloted simulations ((Van der Kamp et. Al, 2005, Pavel, 2010). In a general non-linear 6-dof model the helicopter motion is represented by three translations and three rotations around the body axes-system centred on the helicopter centre of gravity, see Figure A 1 in Appendix A. The helicopter is modelled by dividing it into main components (front rotor, rear rotor, fuselage, horizontal stabilizer, vertical fin) and summing up the contribution of each part to the general system of forces and moments. The following main assumptions were made: 1) Aerodynamic forces and moments are calculated using the blade element theory and integrating along the radius and azimuth to obtain their average effect; 2) The fuselage is modelled with linear aerodynamics; 3) Rotor disc-tilt dynamics (the so-called 'flapping dynamics') are neglected and only steady-state rotor disc-tilt motion is considered; 4) The dynamic inflows of both front and rear rotors are included in the model as state variables and can be described as a quasi-steady dynamic inflow by means of time constants of a value 0.1 sec; 5) The rotors are modelled with a centrally flapping hinge; 6) There are no pitch-flap or pitch-lag couplings; 7) The lead-lag motion of the blades is neglected; 8) The blades are rectangular; 9) There are no tip losses; 10) Gravitational forces are small compared to aerodynamic, inertial and centrifugal forces; 11) The flapping and flow angles are small; 12) The front rotor angular velocity is constant and anticlockwise, the rear rotor angular velocity is constant and clockwise; 13) No reverse flow regions are considered; 14) The flow is incompressible; 15) The blades have a uniform mass distribution; 16) The blade elastic axis, aerodynamic axis, control axis and centre of mass axis coincide; 17) The blades are linearly pre-twisted θ_{twist}= - 9.14 deg. For the expressions of the forces and moments acting on the helicopter components, the reader is referred to APPENDIX A. The equations of motion describing the motion of the helicopter in the 6-dof model are presented in APPENDIX A. To this model, a 6-dof model for the load has been added. The helicopter has three suspension points i=1,2,3 underneath its floor as seen in Figure A6, the tension force in one cable j being Sij. The slings are modelled as linear springs without mass and it is assumed that they have small internal damping. The damping coefficient is set large enough to prevent the oscillating load from inducing a pendulum-like motion. In APPENDIX A more detail is given on the calculation of load sling forces (see eqns. (24) and (25)). The load is modelled as a 6-dof body, the helicopter slings being connected to the load as seen in Figure A7. The slings are modelled as linear weightless springs with a small internal damping. Appendix B presents the Chinook data for the helicopter and load.

3.2 Developing control laws for the longitudinal AFCS

Next, AFCS control laws for operating the Chinook helicopter are added to the presented flight mechanics model. In a classical AFCS, the computation of signals for stabilization, control decoupling and automatic control of the helicopter take place. Generally, an AFCS comprises the sensors, the filters, the stability and control augmentation system (SCAS) and the autopilot which can completely take over the pilot in order to execute certain control tasks.

Fig. 7 (Reijm, Pavel & Bart, 2006) displays for example the pitch control laws connected to the generic helicopter model. Generally, the pitch control laws have 6 input paths: pitch attitude, pitch rate, yaw rate, bank angle, limited airspeed and the longitudinal control position. These signals arrive from the gyros, CPT and pressure sensor and are manipulated to steer the pitch ILCA extensible links to provide pitch rate damping and the DASH actuator to maintain airspeed and a positive stick gradient. The actuation system is quite a complicated mechanism with its own feedback control designed to ensure that the response and stability to control inputs has good performance. Helicopters fitted with an AFCS usually incorporate a limited authority series SCAS actuators, The SCAS authority, i.e. the amount in which a SCAS can overrule the pilot is limited to amplitudes of ±10% of the full actuator throw. For this work, it is assumed that each actuation element can be represented as a first order lag, although this assumption is a crude approximation of the complex behaviour of the servo-elastic system.

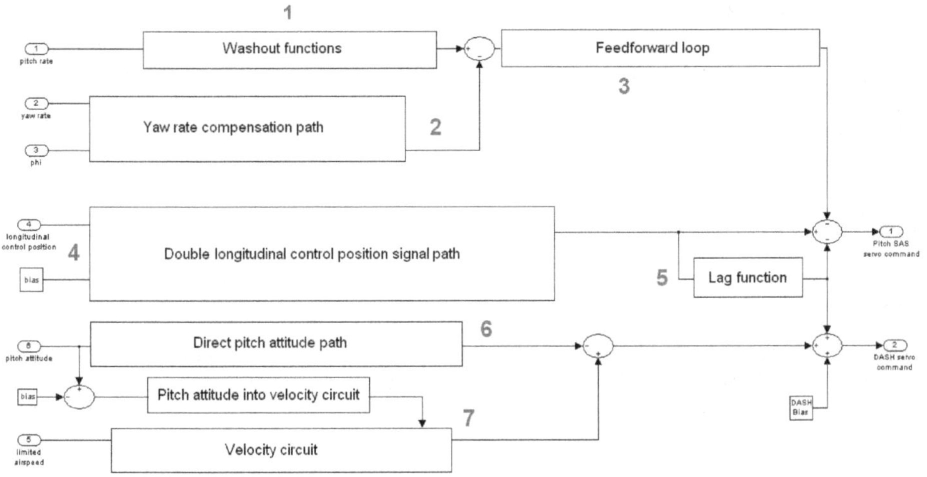

Fig. 7. Pitch control laws for the longitudinal AFCS (Reijm, Pavel & Bart, 2006)

Looking at Fig. 7, one can distinguish the various parts of the control laws. Starting with location 1, this is the primary input for the ILCA path: the pitch rate. It passes through washout functions which act like a sink and filter the signal for certain frequencies. These washout transfer functions can be tuned to select a desired filter strength and a frequency that should be eliminated from the signal. Moving to location 2, the filtered pitch rate signal is compensated by the yaw rate and roll attitude, presumable to cater for the effects of the coupling between the longitudinal and lateral flight dynamics. In location 3 the compensated signal passes through a feedforward loop; the combination of a feedback and feedforward loop increase the responsiveness of a system, decreasing the time to return to the desired state after a disturbance. The longitudinal control position (location 4 in the figure) also contributes to the ILCA signal with the purpose of control augmentation. The CPT signals pass via a direct path and high gain path, which incorporates a *rate limiter*. The rate limiter constricts the appliance of the CPT signal to repress the effects of rapid stick movement, which could otherwise result in overreaction of the system. The direct and rate limited signals are summed and together form the resulting longitudinal CPT signal to be used for the ILCA and DASH actuator. Before the CPT signal reaches the ILCA, it is modified by a special construction involving the regular signal subtracted by the signal that has been filtered by a *lag function*. Generally, the task of a lag function is to cancel high frequency throughput to prevent saturation. They are defined by the transfer function $\frac{1}{\tau s+1}$ in which τ is the time constant that defines the effectiveness of the filter. The higher the time constant, the better is the capacity of filtering high frequencies. The time constant τ is typically between 25 and 100ms, giving actuation bandwidths between 40 and 10 rad/s/. For this study a time constant of 100ms was chosen, resulting in a system with lower actuation bandwidth which shows that the actuation inhibits pilot rapid control actions. Then at location 5 one can see the *washout circuit* designed to only produce output during the transient period, cancelling the steady-state throughput. This behaviour is exactly what is required for control augmentation: the assistance by the boost actuators should be generated only when the pilot is moving the stick, not when keeping it steady. The resulting longitudinal CPT signal not only directs ILCA extension, but also controls the DASH actuator. The signal is supplemented with the pitch attitude that enables pitch attitude hold (location 6). The pitch attitude also works as input for the velocity control circuit in location 7. This velocity control circuit enables the DASH system to provide a positive stick gradient, but also includes feedback and feedforward loops to act on attitude changes caused by gusts.

4. Simulating front cable failure scenarios

The generic control scheme of Fig. 7 is next implemented in the generic tandem helicopter + load model.

The generic helicopter/slung load model as presented above is used next to simulate failure scenarios of the front cable. A tandem V-shape suspended load is assumed to hang underneath the helicopter so that failure of the front hook means actually failure of the two front cables.

In analysing the pilot recovery chances after a front cable failure without dropping the external load, the following question arose: what happens if the remaining cables are not strong enough to carry the load and they snap so that the pilot looses the load. Therefore the tension in the remaining cables was analysed to determine the cases in which the pilot would loose the load. After a cable failure the tension in the remaining cables increases rapidly and may cause them to snap. The snapping of the other cables may cause extra piloting problems, even loss of control and a possible destruction of the rotor above. Generally, generic sling specifications demand a guaranteed life of four years. At the Beginning Of Life (BOL) they must withstand 7 times the designed tensile strength. At the end of their service or End Of Life (EOL) this number has dropped to a minimum of 4.2

times design strength, which is equal to the design strength of suspension points on the external loads. For the Chinook, the tensile strength are calculated with a design load of 17000lb (7.6x104 N) for each cable/hook and 25000lb (11.1x104N) for a combined fore and aft cable/hook (31). During the failure simulations it became clear that the tension in the cables is very high when recovering to normal flight. On many occasions the EOL strength was exceeded and sometimes the BOL strength as well. As it was difficult to decide when a sling reaches its ultimate/breaking strength and snaps, the actual snapping point was set to the BOL strength as this represented a worst-case scenario.

First, to fly the helicopter/slung load system in the front cable failure scenario, simple PID controllers have been developed to generate all control positions. Generally, in a PID controller, it is considered that the collective controls the altitude, the longitudinal cyclic controls pitch attitude, the lateral cyclic controls the roll motion and the pedal controls the sideslip, i.e.:

$$coll = K_z\left(z_{req} - z\right) + K_{\text{int}\_z}\int\left(z_{req} - z\right)dt + K_w w$$

$$long = K_\theta\left(\theta_{req} - \theta\right) + K_{\text{int}\_\theta}\int\left(\theta_{req} - \theta\right)dt + K_q q \tag{1}$$

$$lat = K_\phi\left(\phi_{req} - \phi\right) + K_{\text{int}\_\phi}\int\left(\phi_{req} - \phi\right)dt + K_p p$$

$$pedal = -K_\psi \beta_{fus} + K_{\text{int}\_\psi}\int\left(-\beta_{fus}\right)dt - K_r r$$

At the guidance level, the required pitch attitude is controlled by an altitude hold controller and the required roll angle is controlled by a lateral position hold controller.

$$\theta_{req} = K_h\left(h_{req} - h\right) + K_{\text{int}\_h}\int\left(h_{req} - h\right)dt + K_{hdot}\dot{h} \tag{2}$$

$$\phi_{req} = K_y\left(y_{req} - y\right) + K_{\text{int}\_y}\int\left(y_{req} - y\right)dt + K_{ydot}\dot{y}$$

The longitudinal and lateral pilot inputs are mixed at the level of swashplate as given in (Ostroff, Downing and Rood, 1976).

$$\begin{aligned}
\theta_{1cf} &= \theta_{1cf,pilot}\cdot\cos\beta_{fus} + \theta_{1sf,pilot}\cdot\sin\beta_{fus}\\
\theta_{1sf} &= \theta_{1cf,pilot}\cdot\sin\beta_{fus} - \theta_{1sf,pilot}\cdot\cos\beta_{fus}\\
\theta_{1cr} &= \theta_{1cr,pilot}\cdot\cos\beta_{fus} + \theta_{1sr,pilot}\cdot\sin\beta_{fus}\\
\theta_{1sr} &= -\theta_{1cr,pilot}\cdot\sin\beta_{fus} + \theta_{1sr,pilot}\cdot\cos\beta_{fus}
\end{aligned} \tag{3}$$

Next, the initial PID controlled model is replaced with the longitudinal AFCS controlled system of Fig. 7, the other inputs remaining still PID controlled. This new pilot model is used to simulate different failure scenarios of the front cable. The following steps are taken to complete a failure scenario with the longitudinal AFCS switched on:

- Calculate the trim state for a desired airspeed;
- Use Bilinear Transformation to map the continuous time transfer functions to the discrete domain;

- Compute the new pilot input value for smooth transition;
- Simulate the first 25 seconds with a full PID control. This is done because when the AFCS is turned on in flight after it has been switched off, or it leaves ground mode, a DASH error signal could exist. This means that DASH actuator operation and the longitudinal stick position are out of sync, requiring a small period of time for the actuator to cancel the error signal after which it is able to resume normal operation. This is done with the Transition Mode of the DASH. Discussions with RNLAF test pilots provided the indication that approximately 25 seconds is a realistic value;
- At 25 seconds, switch to longitudinal AFCS with all other inputs (collective, lateral and pedal) PID controlled;
- At t = 26 seconds, a cable failure will be introduced. The control laws will keep operating, immediately responding to the changing helicopter states.
- From the start of the failure until the specified pilot reaction time of either 0, 1 or 2 seconds, the PID controllers will enter a stick-fixed mode, remaining at the position they were in at t = tfailure.
- When the reaction time period has expired, the PID controllers will kick in and the simulation will again run in mixed mode until it is finished, covering the same 5 seconds from t = tfailure.

To determine the limits within which the pilot can control the recovery, the ADS-33 standard (ADS-33, 2000), section 3.1.14.4 'Transients Following Failures' has been used. According to ADS-33, for rotorcraft without failure warning and cueing devices, the perturbations encountered shall not exceed the limits as given in Table 1 (equivalent with Table III from ADS-33). Also, according to pilot experience it appeared that the pilot can react within 1 or 2 seconds after the load failure. Therefore, the pilot reaction time in the simulations performed in this paper has been adjusted to 1 and maximum 2 seconds.

Level	Flight condition		
	Hover and Low speed	Forward Flight	
		Near-Earth	Up-and-away
1	3° roll, pitch, yaw 0.05g n_x, n_y, n_z No recovery action for 3.0 sec	Both Hover and Low Speed and Forward Flight Up-and-Away requirements apply	Stay within OFE. No recovery action for 10 sec
2	10° attitude change 0.2g acceleration No recovery action for 3.0 sec	Both Hover and Low Speed and Forward Flight Up-and-Away requirements apply	Stay within OFE. No recovery action for 5.0 sec
3	24° attitude change 0.4g acceleration No recovery action for 3.0 sec	Both Hover and Low Speed and Forward Flight Up-and-Away requirements apply	Stay within OFE. No recovery action for 10 sec

Table 1. Transients following failures (ADS-33, 2000)

A typical failure simulation can be seen in Fig. 8. The case considered is the helicopter flying forward at 50 kts when a front failure occurs with a 2-point suspended load of 2 tonnes. It is considered that the pilot reaction starts 1 sec after the failure. The controls remain unchanged from the moment of failure to the moment of pilot reaction (i.e. for 1 second). Looking at this figure it appears that the pilot can control this failure, being close to the upper limit for the longitudinal controller and well within the limits for the collective and lateral controllers.

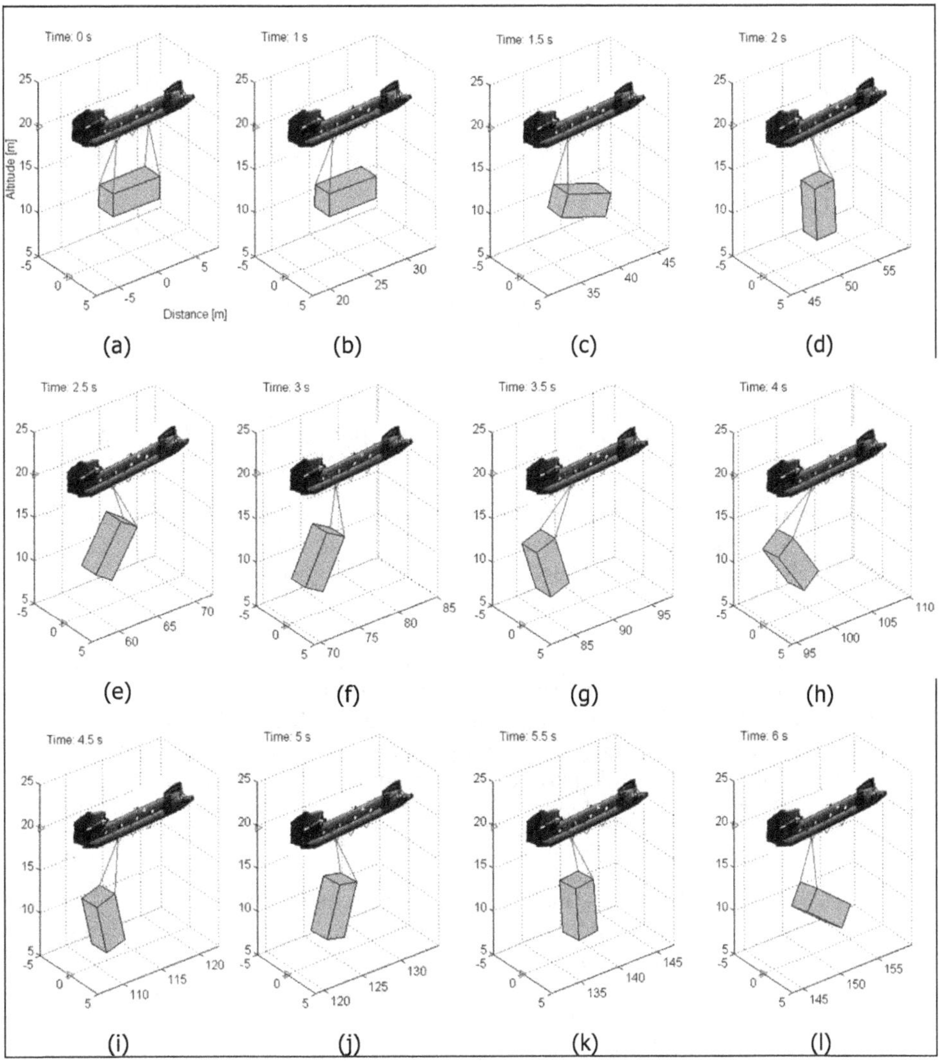

Fig. 8. Helicopter and load snapshot to a front cable failure scenario from 50 kts initial velocity, 2-point suspension, 2000 kg load, 1 sec delay in pilot reaction

Concerning the time responses of the servo commands control actions, Fig. 9 displays the pitch control law of the actuator servo commands during a 30-second level flight at 50 kts.

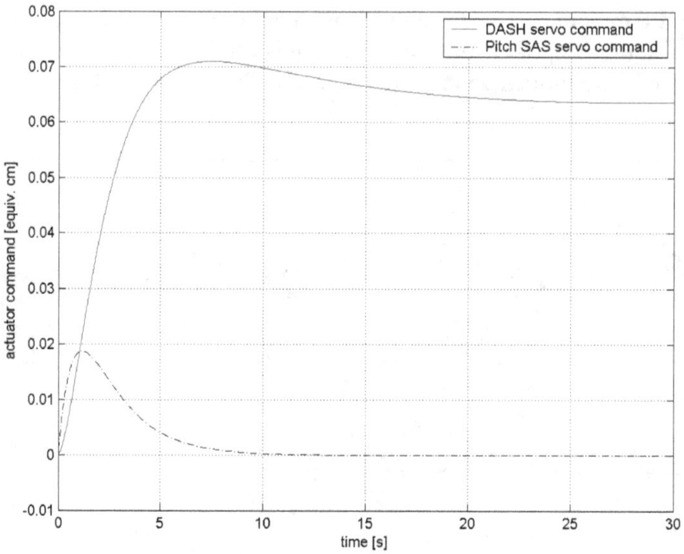

Fig. 9. Pitch control low for the DASH and SAS actuator command during the 50 kts forward flight

Looking at this figure, one can see that there is a 25-second settling time that the DASH actuator requires before reaching a steady state value of about 6.3 cm.. Such behaviour seems to agree with the RNLAF CH-47D flight manual ((Anon, 2004),section 8.3). The manual states that when the AFCS is turned on in flight after it has been switched off, or it leaves ground mode, a DASH error signal could exist. This means that DASH actuator operation and the longitudinal stick position are out of sync, requiring a small period of time for the actuator to cancel the error signal after which it is able to resume normal operation. This is done with the Transition mode of the DASH as described above. The manual does not mention how long this period actually lasts, but, from discussions with Chinook pilots, it appears that this period it would amount to approximately 25 seconds, value similar to the one encountered in Fig. 9.

Furthermore, the following scenarios are considered:

- the load is suspended on two-point or three-point suspension systems. In the three-point suspension the redundant set of slings (between the front and rear hook) come into action when the front cables fail.
- the load can be suspended either in a nose-down position (so-called 'nose-down rigged load') or in a horizontal position (so-called 'level rigged load').
- the helicopter is flying initially in level flight at velocities varying between 10 and 100 kts
- the helicopter can carry three different container of 2 tonnes, 6 tonnes or 10 tonnes.
- the pilot reaction time to failure varies from instantaneously reaction (ideal case) to a delay in response of 1 and 2 seconds.

Fig. 10 shows the helicopter pitch rate, attitude and velocities after a cable failure in the case of a PID controlled helicopter (left hand side of the figure) and an AFCS controlled helicopter (right hand side). The case considered is the helicopter flying forward at 50 kts when a front failure occurs with a 2-point suspended load of 2000 kg. It is considered that the pilot PID control will start to react at 1 sec after the front suspension point has failed. The controls remain unchanged from the moment of failure to the moment of pilot reaction (i.e. for 1 second). It is interesting to discuss how the AFCS affects the simulation. While the PID controller is restrained by the pilot reaction time, the AFCS control laws immediately start reacting to the change in helicopter pitch attitude caused by the swinging of the load. This should result in a lower maximum pitch rate, an expectation confirmed by the results. The maximum pitch rate q_{max} of the PID controlled helicopter is equal to 8.6 deg/sec (27th second), whereas the q_{max} resulting from AFCS control comes to 7.6 deg/sec (2nd second), an 11% attitude reduction. Concerning the pitch attitude, one can see that the AFCS controller clearly has trouble to maintain the trimmed pitch attitude, allowing it to increase to 8 degrees. This immediately has its effect on the forward velocity that begins to decline as soon as the pitch angle starts to build up at about 27 seconds. The filtering of the pitch attitude control also has its effect on the vertical velocity, but this is neither vital to the survival of the helicopter nor relevant to the ADS-33 handling limits. Simulating different scenarios it was observed that the DASH control circuit fails to maintain airspeed when flying faster than 40 kts. Though it appears more and more likely that the implementation of the DASH control circuit does not wholly agree with the real-life situation, the pitch ILCA is still doing its job just as intended, keeping the pitch rate within handling limits and making recovery possible.

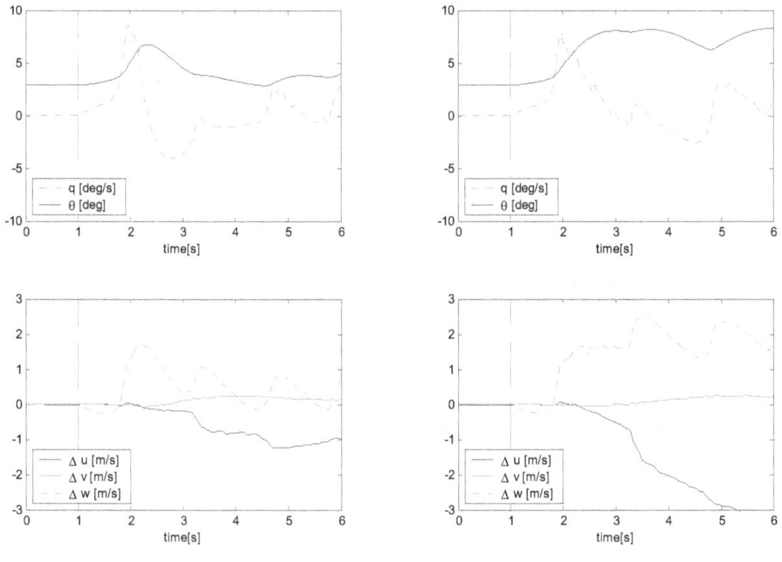

PID controlled AFCS controlled

Fig. 10. Helicopter motion after front suspension load failure, 50kts, two-point suspension, 2 tonne mass load

Fig. 10 clearly shows the advantage of the running AFCS: the pitch ILCA picks up on the increasing pitch rate right after the cable failure, resulting in a lower maximum pitch rate compared to PID control with a one second reaction time delay. The PID controlled pitch rate ranges from -4 to +8 deg/s, the AFCS improves on this with a range of -2.5 to +7.5 deg/s.

The difference in helicopter motions result in a small difference in the way the load swings. But surprisingly, the change is not to be found in the axis in which the controller actually differs. Fig. 11 presents the pitch, roll and yaw motions of the container. Apart from the deviation in shape, the longitudinal AFCS controller causes an increased negative load roll rate and larger overall yaw rate amplitude. The swinging of the load may even result in a destructive collision between the container and helicopter hull before the load can be safely detached. This danger should be investigated in future research.

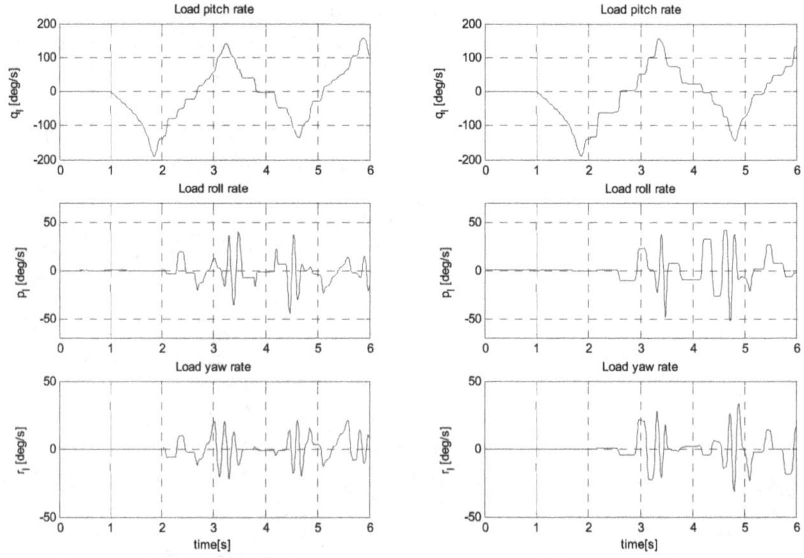

PID controlled AFCS controlled

Fig. 11. Load motion after front suspension cable failure

5. Defining safety envelopes for the load failures

The numerous failure scenarios simulated have been plotted in safety envelopes giving the velocity when recovery was possible as a function of load mass carried. Fig. 12 and Fig. 13 present the envelopes for a two- and respectively three-point suspension level rigged container after the front suspension point failed as a function of pilot reaction time to recover. Two controls are analysed: (a) a PID controller and (b) an AFCS controller. Looking at Fig. 12 one can see that in case of a two-point suspension, the PID control covers a larger envelope area than the partial AFCS control when the pilot reacts instantaneously to the failure (τ= 0 sec). However, in case of a three-point suspension this difference is not present. Fig. 12(b) shows a kind of inverse trend in safety when switching on the AFCS: if the pilot reacts instantaneously to the failure (τ=0 dotted line envelope) he has less chances to recover than if he reacts one second later (τ =1sec point dotted line). This can be explained as follows: the AFCS is actually first attempting to converge to reference (trim) values. If the pilot reacts immediately to the failure, the AFCS has difficulties in combining pilot inputs with its own convergence algorithm. But as soon as the circumstances favour the AFCS controller, i.e. the pilot reaction time becomes 1 or 2 seconds, the advantage of its application becomes clear.

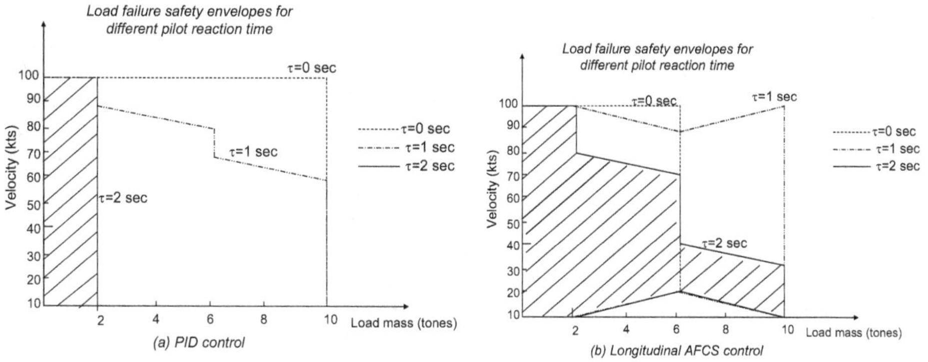

Fig. 12. Safety envelopes for a two-point suspension

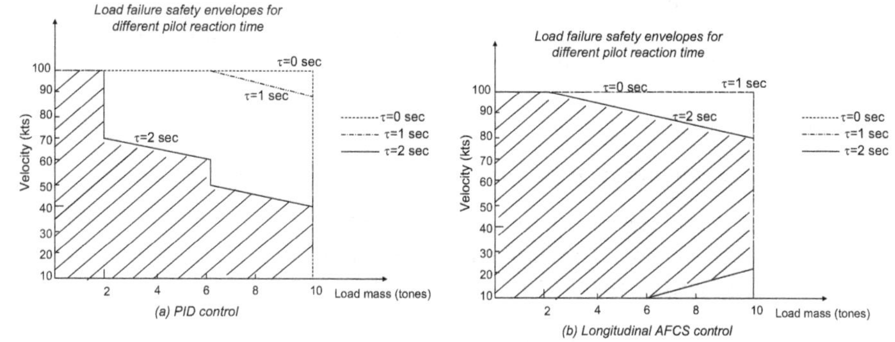

Fig. 13. Safety envelopes for a three-point suspension

Especially with a 2 second reaction time the contribution of the AFCS to the helicopter recovery emerges as invaluable. This can be seen when comparing Fig. 12(a) with Fig. 12(b), showing even more than doubling the envelope area when τ=2 sec and when instead of using PID controller the AFCS is switched on. This means that, when compared to a basic PID controlled helicopter, the AFCS is offering a much broader flight regime in which under slung loads can be transported. While the PID only stabilizes the helicopter motions by limiting the rate at which the attitude changes, the AFCS is attempting to converge to trim values.

Comparing the two-point and three-point suspension for an AFCS controlled helicopter (i.e. Fig. 12(b) with Fig. 13(b)) it appears that a three-point suspension is safer. However, this does not mean that the two-point suspension is not safe for a certain condition. For example, if a 6 tone load needs to be transported, one can choose for a two-point suspension with the condition of not exceeding 70 kts in level flight. Above this velocity the chance to recover in case of a load failure is questionable. In case of a 10 tones load, the AFCS for a two-point suspension cannot contribute to the pilot efforts to recover the load if the failure occurs above 30 kts. In this case it is safer to choose a three-point suspension for load transportation.

6. Conclusion

The purpose of this chapter is to determine how the AFCS influenced the recovery prospects of a Chinook helicopter with an external sling load when one of its cables brakes. This can be of vital importance for deciding whether to replace the safer three-point redundant suspension to a two-point suspension during Royal Netherlands Air Force operations. The chapter presented the flight control laws for a longitudinal automatic flight control system (AFCS). The general conclusion that could be drawn was that the assistance of the AFCS expanded the number of occasions at which the helicopter motions stayed within the handling limits during recovery of a cable failure. In many cases, the AFCS cancelled the negative effect of a delayed pilot response, supporting the supposition that the combination of active pilot control and AFCS backup could push the flight envelope boundaries even further. The chapter proposed safety envelopes covering the areas that define the conditions at which a cable failure at a suspension point could happen without presumable fatal consequences. It was demonstrated that since the longitudinal part of the AFCS was operating continuously, it was not bound by the obstruction of the imposed pilot reaction time. For this reason, partial AFCS control gained an advantage when the pilot reaction was delayed. Flying with a two-point suspension is just as safe as a three-point suspension stipulating that the safety envelope boundaries are obeyed.

7. Appendix A

7.1 Derivation of the equations of motion in a six degree-of-freedom model

In a general 6-dof non-linear body model the helicopter motion is represented by three translations and three rotations around the body axes-system $\overline{E}_B\{X_B \ \ Y_B \ \ Z_B\}$ centred in the helicopter centre of gravity, see Fig. A1.

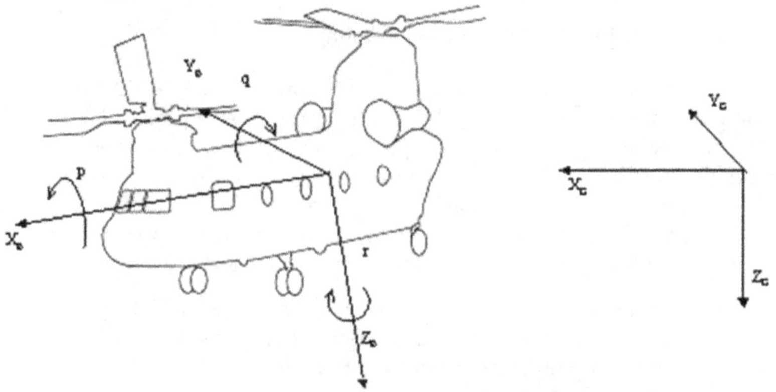

Fig. A1. System of coordinates used to express the helicopter motion in the 6-dof model

The system of equations describing completely the motion of the helicopter in an inertial system are:

$$\dot{u} = -g\sin\theta + X/_{m} + rv - qw$$
$$\dot{v} = g\sin\phi\cos\theta + Y/_{m} - ru + pw$$
$$\dot{w} = g\cos\phi\cos\theta + Z/_{m} + qu - pv$$

$$\dot{p} = \frac{rq\left(I_yI_z - I_z^2 - I_{xz}^2\right) - pqI_{xy}\left(I_x - I_y + I_z\right) + I_zL + I_{xz}N}{I_xI_z - I_{xz}^2}$$
$$\dot{q} = \frac{M + pr\left(I_z - I_x\right) + I_{xz}\left(r^2 - p^2\right)}{I_y} \tag{4}$$
$$\dot{r} = \frac{-rqI_{xz}\left(I_x - I_y + I_z\right) + pq\left(I_x^2 - I_xI_y + I_{xz}^2\right) + I_{xz}L + I_xN}{I_xI_z - I_{xz}^2}$$

$$\dot{\psi} = \frac{\left(q\sin\phi + r\cos\phi\right)}{\cos\theta}$$
$$\dot{\theta} = q\cos\phi - r\sin\phi$$
$$\dot{\phi} = p + \dot{\psi}\sin\theta$$

In order to describe completely the helicopter motion w.r.t. the Earth system, the equations of trajectory can be added:

$$\dot{x} = \left(u\cos\theta + \left(v\sin\phi + w\cos\phi\right)\sin\theta\right)\cos\psi - \left(v\cos\phi - w\sin\phi\right)\sin\psi$$
$$\dot{y} = \left(u\cos\theta + \left(v\sin\phi + w\cos\phi\right)\sin\theta\right)\sin\psi + \left(v\cos\phi - w\sin\phi\right)\cos\psi \tag{5}$$
$$\dot{z} = -u\sin\theta + \left(v\sin\phi + w\cos\phi\right)\cos\theta$$

To these systems of equations, two differential equations are added for the dynamic inflow of the front main and rear rotors, describing the dynamic inflow as a "quasi-steady inflow" by means of time constants:

$$\tau_f\dot{\lambda}_{if} = C_{T,Elem,f} - C_{T,Glau,f}$$
$$\tau_r\dot{\lambda}_{ir} = C_{T,Elem,r} - C_{T,Glau,r} \tag{6}$$

where $C_{T,elem}$ and $C_{T,Glau}$ are the rotor thrust coefficients as expressed in blade element theory and respectively Glauert theory. The total forces and moments acting on the helicopter centre of gravity consist of the sum of front and rear rotors (with indices Rf and Rr), rotor torque (Q) and helicopter body aerodynamics (B). For the load, a suspension point component (S) is considered which is added when the load is attached underneath the helicopter model.

$$
\begin{aligned}
X &= X_{Rf} + X_{Rr} + X_B(+X_S) \\
Y &= Y_{Rf} + Y_{Rr} + Y_B(+Y_S) \\
Z &= Z_{Rf} + Z_{Rr} + Z_B(+Z_S) \\
L &= L_{Qf} + L_{Qr} + L_{Rf} + L_{Rr} + L_B(+L_S) \\
M &= M_{Rf} + M_{Rr} + M_B(+M_S) \\
N &= N_{Qf} + N_{Qr} + N_{Rf} + N_{Rr} + N_B(+N_S)
\end{aligned}
\tag{7}
$$

The rotor forces and moments consist of front and rear vertical thrust components T_f and T_r, drag forces H_{Xf} and $H_{X,r}$, lateral forces H_{Yf}, and H_{Yr} and rotor shaft torques M_{Qf} and M_{Qr} as seen in Fig. A2.

$$
\begin{aligned}
X_{Rf} &= X_{Tf} + X_{Hf} & \qquad X_{Rr} &= X_{Tr} + X_{Hr} \\
Y_{Rf} &= Y_{Tf} + Y_{Hf} & \qquad Y_{Rr} &= Y_{Tr} + Y_{Hr} \\
Z_{Rf} &= Z_{Tf} + Z_{Hf} & \qquad Z_{Rr} &= Z_{Tr} + Z_{Hr}
\end{aligned}
\tag{8}
$$

$$
\begin{aligned}
L_{Rf} &= Y_{Rf}h_f - Z_{Rf}f_1 + L_{Qf} & \qquad L_{Rr} &= Y_{Rr}h_r - Z_{Rr}f_1 + L_{Qr} \\
M_{Rf} &= -X_{Rf}h_f - Z_{Rf}l_f & \qquad M_{Rr} &= -X_{Rr}h_r + Z_{Rr}l_r \\
N_{Rf} &= X_{Rf}f_1 + Y_{Rf}l_f + N_{Qf} & \qquad N_{Rr} &= X_{Rr}f_1 - Y_{Rr}l_r + N_{Qr}
\end{aligned}
$$

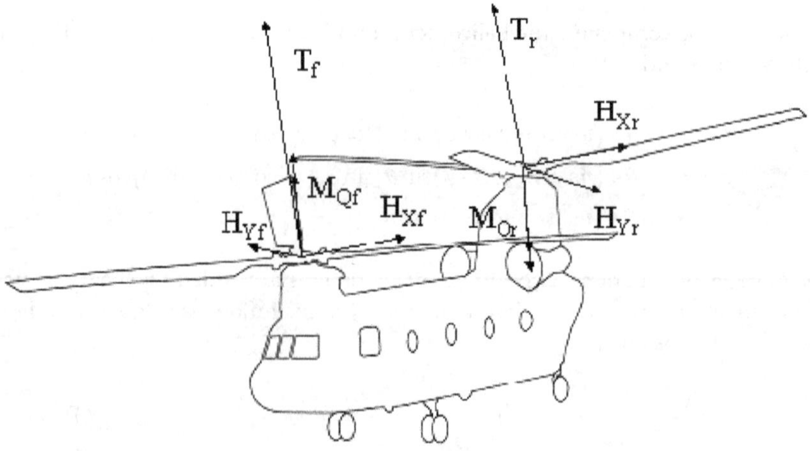

Fig. A2. Forces and Moments on the rotors

The model uses the following sign conventions (see Fig. A3 and Fig. A4): 1) Longitudinal disc tilt for front and rear rotors a_{1f} and a_{1r} are assumed positive for backward tilted rotor disc plane; 2) Lateral disc tilt for front and rear rotor b_{1f} and b_{1r} are positive for rotor disc plane tilted in the direction of azimuth angle ψ=90°(i.e. $b_{1f}>0$ to the right and $b_{1r}>0$ to the left, backside view); 3) Collective pitch for front and rear rotor θ_{0f} and θ_{0r} are positive when the pilot moves the collective up; 4) Longitudinal cyclic for front and rear rotor θ_{1sf} and θ_{1sr} are assumed positive when the pilot moves the stick forward; 5) Lateral cyclic for front and rear rotors θ_{1cf} and θ_{1cr} are assumed positive when the pilot moves the stick to the right for cyclic pitch to the right.

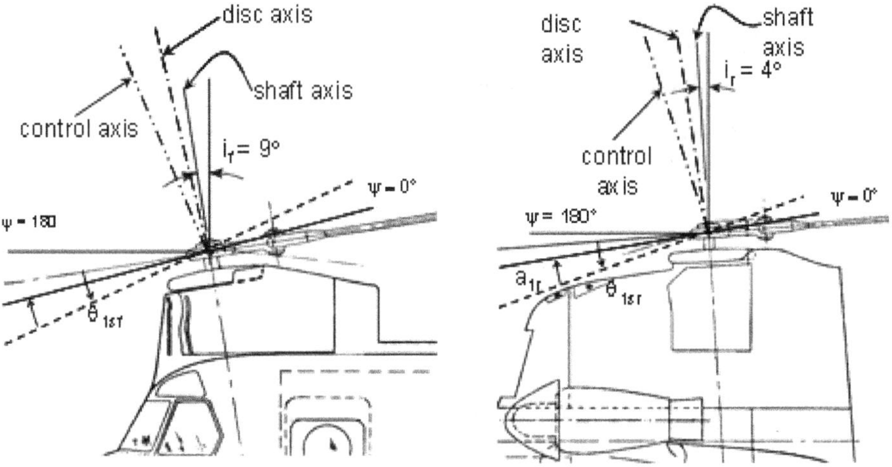

Fig. A3. Longitudinal axis Chinook control (side view)

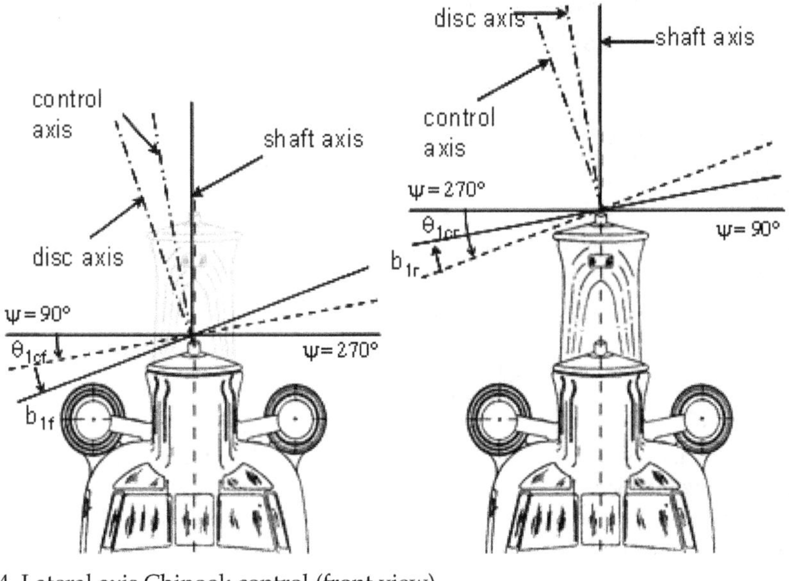

Fig. A4. Lateral axis Chinook control (front view)

The contributions of these components to the front rotor thrust and the horizontal and side forces in the helicopter body system are:

$$
\begin{aligned}
X_{Tf} &= T_f \sin\left(\theta_{1sf} - a_{1f} + i_f\right)\cos\left(\theta_{1cf} + b_{1f}\right) \\
X_{Hf} &= -H_{Xf}\cos\left(\theta_{1sf} - a_{1f} + i_f\right) - H_{Yf}\sin\left(\theta_{1cf} + b_{1f}\right)\sin\left(\theta_{1sf} - a_{1f} + i_f\right) \\
Y_{Tf} &= T_f \sin\left(\theta_{1cf} + b_{1f}\right)\cos\left(\theta_{1sf} - a_{1f} + i_f\right) \\
Y_{Hf} &= H_{Yf}\cos\left(\theta_{1cf} + b_{1f}\right) \\
Z_{Tf} &= -T_f \cos\left(\theta_{1sf} - a_{1f} + i_f\right)\cos\left(\theta_{1cf} + b_{1f}\right) \\
Z_{Hf} &= -H_{Xf}\sin\left(\theta_{1sf} - a_{1f} + i_f\right) + H_{Yf}\sin\left(\theta_{1cf} + b_{1f}\right)\sin\left(\theta_{1sf} - a_{1f} + i_f\right) \\
L_{Qf} &= M_{Qf}\sin i_f \\
N_{Qf} &= -M_{Qf}\cos i_f
\end{aligned}
\tag{9}
$$

For the rear rotor, these components are (note the opposite direction of the lateral flapping angle b_{1r} with respect to the front rotor equations):

$$
\begin{aligned}
X_{Tr} &= T_r \sin\left(\theta_{1sr} - a_{1r} + i_r\right)\cos\left(\theta_{1cr} - b_{1r}\right) \\
X_{Hr} &= -H_{Xr}\cos\left(\theta_{1sr} - a_{1r} + i_r\right) + H_{Yr}\sin\left(\theta_{1cr} - b_{1r}\right)\sin\left(\theta_{1sr} - a_{1r} + i_r\right) \\
Y_{Tr} &= T_r \sin\left(\theta_{1cr} - b_{1r}\right)\cos\left(\theta_{1sr} - a_{1r} + i_r\right) \\
Y_{Hr} &= -H_{Yr}\cos\left(\theta_{1cr} - b_{1r}\right) \\
Z_{Tr} &= -T_r \cos\left(\theta_{1sr} - a_{1r} + i_r\right)\cos\left(\theta_{1cr} - b_{1r}\right) \\
Z_{Hr} &= -H_{Xr}\sin\left(\theta_{1sr} - a_{1r} + i_r\right) - H_{Yr}\sin\left(\theta_{1cr} - b_{1r}\right)\cos\left(\theta_{1sr} - a_{1r} + i_r\right) \\
L_{Qr} &= -M_{Qr}\sin i_r \\
N_{Qr} &= M_{Qr}\cos i_r
\end{aligned}
\tag{10}
$$

The front and the rear rotor thrust (T_f and T_r), horizontal force (drag force) (H_{Xf}, H_{Xr}), lateral forces (H_{Yf}, H_{Yr}) and torque (M_{Qf}, M_{Qr}) are expressed by their non-dimensional coefficients:

$$
\begin{aligned}
T_{f,r} &= C_{Tf,r}\,\rho(\Omega R)^2 \pi R^2 \\
H_{Xf,r} &= C_{HXf,r}\,\rho(\Omega R)^2 \pi R^2 \\
H_{Yf,r} &= C_{HYf,r}\,\rho(\Omega R)^2 \pi R^2 \\
M_{Qf,r} &= C_{Qf,r}\,\rho(\Omega R)^2 \pi R^3
\end{aligned}
\tag{11}
$$

The non-dimensional coefficients for rotor thrust, horizontal and lateral forces and torque can be calculated using the blade element theory by integration of the lift and drag forces on each blade element along the blade and around the azimuth. Their non-dimensional coefficients w.r.t. the disc plane (non-dimensionalized by $\rho(\Omega R)^2(\pi R^2)$), for the front and rear rotors, are:

$$C_{Tf,r} \approx C_{T,elem,f,r} = \frac{\sigma C_{l\alpha}}{2}\left[\frac{\theta_{0f,r}}{3}\left(1+\frac{3}{2}\mu_{xf,r}\right)+\frac{\mu_{zf,r}-\lambda_{if,r}}{2}\right] \tag{12}$$

$$\begin{aligned} C_{HXf,r} = \frac{\sigma C_{l\alpha}}{2}\left[\frac{\mu_{xf,r}C_{df,r}}{2C_{l\alpha}}+\theta_{0f,r}\left(\frac{a_{1f,r}}{3}-\frac{\mu_{xf,r}}{2}\frac{\mu_{zf,r}-\lambda_{if,r}}{2}\right)+\right. \\ \left.+\frac{3(\mu_{zf,r}-\lambda_{if,r})a_{1f,r}}{2}+\frac{\mu_{xf,r}}{4}\left(a_{0f,r}{}^{2}+a_{1f,r}{}^{2}\right)-\frac{a_{0f,r}b_{1f,r}}{6}\right]-C_{Tf,r}a_{1f,r} \end{aligned} \tag{13}$$

$$\begin{aligned} C_{HYf,r} = \frac{\sigma C_{l\alpha}}{2}\left[\mu_{xf,r}{}^{2}\left(\frac{b_{1f,r}\theta_{0f,r}}{2}-a_{0f,r}a_{1f,r}\right)+\mu_{xf,r}\cdot\left(\frac{a_{1f,r}b_{1f,r}}{4}+\frac{3\left(\mu_{zf,r}-\lambda_{if,r}\right)a_{0f,r}}{2}-\frac{3a_{0f,r}\theta_{0f,r}}{4}\right)+\right. \\ \left.+\frac{\theta_{0f,r}b_{1f,r}}{3}+\frac{3\left(\mu_{zf,r}-\lambda_{if,r}\right)b_{1f,r}}{4}+\frac{a_{0f,r}a_{1f,r}}{6}\right]-C_{Tf,r}b_{1f,r} \end{aligned} \tag{14}$$

The main rotor thrust coefficient in Glauert theory is:

$$C_{T,Glau} = 2\lambda_{if,r}\sqrt{\mu_{xf,r}{}^{2}+(\lambda_{if,r}-\mu_{zf,r})^{2})} \tag{15}$$

The rotor torque coefficient can be expressed as:

$$C_{Qf,r} = \frac{\sigma C_{df,r}}{8}\left(1+4.7\mu_{xf,r}{}^{2}\right)-C_{Tf,r}\left(\mu_{zf,r}-\lambda_{if,r}\right)-\mu_{xf,r}C_{HXf,r} \tag{16}$$

The advance ratios of front/rear rotors $\mu_{xf,r}$, $\mu_{xf,r}$, $\mu_{xf,r}$ are calculated by projecting helicopter velocity vector (u,v,w) on the plane of front/rear rotor hub as ($u_{f,r}$, $v_{f,r}$, $w_{f,r}$):

$$\begin{bmatrix} u_f \\ v_f \\ w_f \end{bmatrix} = \begin{bmatrix} u+f_1r-h_fq \\ v+l_fr+h_fp \\ w-l_fq-f_1p \end{bmatrix} \quad ; \quad \begin{bmatrix} u_r \\ v_r \\ w_r \end{bmatrix} = \begin{bmatrix} u+f_1r-h_rq \\ v-l_rr+h_rp \\ w+l_rq-f_1p \end{bmatrix} \tag{17}$$

and then by non-dimensionalizing ($u_{f,r}$, $v_{f,r}$, $w_{f,r}$) by ΩR while neglecting the yaw rate r. This leads to the following expressions for the front/rear rotor advance ratios:

$$\mu_{xf} = \frac{\left(u-h_fq\right)}{\Omega R} \qquad \mu_{yf} = \frac{\left(v+h_fp\right)}{\Omega R} \qquad \mu_{zf} = \frac{\left(w-l_fq\right)}{\Omega R}$$

$$\mu_{xr} = \frac{\left(u-h_rq\right)}{\Omega R} \qquad \mu_{yr} = \frac{\left(v+h_rp\right)}{\Omega R} \qquad \mu_{zr} = \frac{\left(w+l_rq\right)}{\Omega R} \tag{18}$$

The blade drag coefficient $C_{df,r}$ of front/rear rotors is calculated as:

$$C_{df,r} = C_{d0}+C_{dt}C_{Tf,r}^{2} \tag{19}$$

with the values C_{d0} and C_{dt} defined in Appendix B. The rotor interference is considered through interference factors expressed as in (Ostroff, Downing and Rood, 1976):

$$dinterf_{\text{f-on-r}} = \left[0.356 + 0.321\chi_{If} - 0.368\chi_{If}^{\ 2} + 0.392\chi_{If}^{\ 3}\right]\left(1 - \left|\sin\beta_{fr}\right|\right) + \\ + \left[0.356 + 0.0131\chi_{If} - 0.0764\chi_{If}^{\ 2} - 0.0085\chi_{If}^{\ 3}\right]\cdot\left|\sin\beta_{fr}\right|$$

$$dinterf_{\text{r-on-f}} = \left[0.356 + 0.151\chi_{Ir} - 0.314\chi_{Ir}^{\ 2} + 0.164\chi_{Ir}^{\ 3}\right]\left(1 - \left|\sin\beta_{rr}\right|\right) + \\ + \left[0.356 + 0.0131\chi_{Ir} - 0.0764\chi_{Ir}^{\ 2} - 0.0085\chi_{Ir}^{\ 3}\right]\cdot\left|\sin\beta_{rr}\right| \tag{20}$$

where χ_{If}, χ_{Ir} are the inflow angles of front and rear rotors defined as in Figure A5 and equation (21) and β_{fr} and β_{rr} are rotor sideslip angles defined in equation (22) and (17).

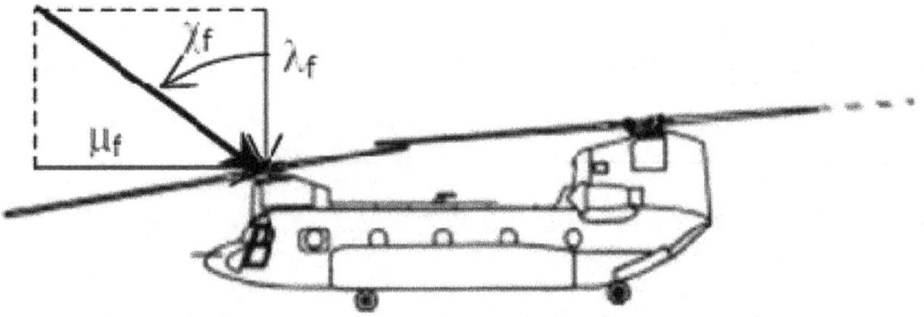

Fig. A5. Rotor inflow angle used to calculate front and rear rotor interferences

$$\chi_{If} = \arctan\left(\frac{\mu_{xf}}{\lambda_f}\right) \qquad \chi_{Ir} = \arctan\left(\frac{\mu_{xr}}{\lambda_r}\right) \tag{21}$$

$$\begin{aligned} \sin\beta_{fr} &= v_f \Big/ \sqrt{u_f^2 + v_f^2} & \cos\beta_{fr} &= u_f \Big/ \sqrt{u_f^2 + v_f^2} & u_f &\neq 0 \\ \sin\beta_{rr} &= v_r \Big/ \sqrt{u_r^2 + v_r^2} & \cos\beta_{rr} &= u_r \Big/ \sqrt{u_r^2 + v_r^2} & u_r &\neq 0 \end{aligned} \tag{22}$$

The interference factors are used to compute a new inflow ratio in the Glauert formula (15):

$$\begin{aligned} \lambda_{if,new} &= \frac{C_{Tf}}{2\sqrt{\left(\mu_{zf} - \lambda_{if}\right)^2 + \mu_{xf}^{\ 2}}} + dinterf_{\text{r-on-f}} \frac{C_{Tr}}{2\sqrt{\left(\mu_{zr} - \lambda_{ir}\right)^2 + \mu_{xr}^{\ 2}}} \\ \lambda_{ir,new} &= \frac{C_{Tr}}{2\sqrt{\left(\mu_{zr} - \lambda_{ir}\right)^2 + \mu_{xr}^{\ 2}}} + dinterf_{\text{fonr}} \frac{C_{Tf}}{2\sqrt{\left(\mu_{zf} - \lambda_{if}\right)^2 + \mu_{xf}^{\ 2}}} \end{aligned} \tag{23}$$

The fuselage forces and moments are calculated through flat plate theory using (Ostroff, Downing and Rood, 1976):

$$
\begin{aligned}
X_B &= -C_{FE}\tfrac{1}{2}\rho V^2 \cos\alpha_{fus}\cos\beta_{fus} \\
Y_B &= -C_{Y\beta}\tfrac{1}{2}\rho V^2 \sin\beta_{fus} \\
Z_B &= -C_{L\alpha\, fus}\tfrac{1}{2}\rho V^2 \sin\alpha_{fus} \\
L_B &= -C_{L\beta}\tfrac{1}{2}\rho V^2 \sin\beta_{fus}\left|\cos\beta_{fus}\right|\left(1-\left|\sin\alpha_{fus}\right|\right) \\
M_B &= C_{M\alpha}\tfrac{1}{2}\rho V^2 \sin\alpha_{fus}\cos\alpha_{fus} \\
N_B &= -C_{N\beta}\tfrac{1}{2}\rho V^2 \sin\beta_{fus}\cos\beta_{fus}\left(0.94\sin\alpha_{fus}+0.342\cos\alpha_{fus}\right)
\end{aligned}
\tag{24}
$$

where the fuselage angle of attack α_{fus} and fuselage sideslip β_{fus} are defined as seen as:

$$
\begin{aligned}
\sin\alpha_{fus} &= w\big/\sqrt{u^2+w^2} \qquad & \cos\alpha_{fus} &= u\big/\sqrt{u^2+w^2} \\
\sin\beta_{fus} &= v\big/\sqrt{u^2+v^2} \qquad & \cos\beta_{fus} &= u\big/\sqrt{u^2+v^2}
\end{aligned}
\tag{25}
$$

The helicopter has three suspension points i=1,2,3 underneath its floor as seen in Fig. A6, the tension force in one cable j being S_{ij}. The total forces on the slings are:

$$
\begin{aligned}
X_S &= S_{ij}\cos\alpha_{ij}\sin\beta_{ij} \\
Y_S &= S_{ij}\sin\alpha_{ij}\sin\beta_{ij} \\
Z_S &= S_{ij}\cos\beta_{ij} \\
L_S &= -S_{ij}\left\{\sin\alpha_{ij}\sin\beta_{ij}hs_i - \cos\beta_{ij}f_1\right\} \\
M_S &= S_{ij}\cos\alpha_{ij}\sin\beta_{ij}hs_i - S_{1,3j}\cos\beta_{1,3j}fs_{1,3} + S_{2j}\cos\beta_{2j}fs_2 \\
N_S &= S_{ij}\cos\alpha_{ij}\sin\beta_{ij}f_1 + S_{1,3j}\sin\alpha_{1,3j}\sin\beta_{1,3j}fs_{1,3} - S_{2j}\sin\alpha_{2j}\sin\beta_{2j}fs_2
\end{aligned}
\tag{26}
$$

where α_{ij} and β_{ij} are the angles that the sling cable makes with the X and Z direction; $fs_{1,2,3}$ and $hs_{1,2,3}$ are the hook distances to the helicopter centre of gravity. The slings are modelled as linear springs without mass and have small internal damping. The tension force in each cable is determined by comparing the actual sling length l_{ij} to the zero tension sling length l_{0ij}, see equation (27), where j is the cable number on the ith suspension point. The zero tension cable length as well as the cable stiffness are known, the actual length of each cable is computed by measuring the distance between the specific helicopter suspension point and the suspension point on the load.

$$
S_{ij} = \max\left\{0, k\left(l_{ij} - l_{0ij}\right)\right\}
\tag{27}
$$

The damping forces inside the slings are determined using equation (28). In this equation dV is the relative velocity between the endpoints of a cable, in the direction of that cable. This requires that both endpoint velocities are known in the same axis system. The damping coefficient is set just large enough to prevent the load from increase in pendulum motion when the load is oscillating around a fixed point.

$$
S_{damp,ij} = dV_{ijl}c_{damp}
\tag{28}
$$

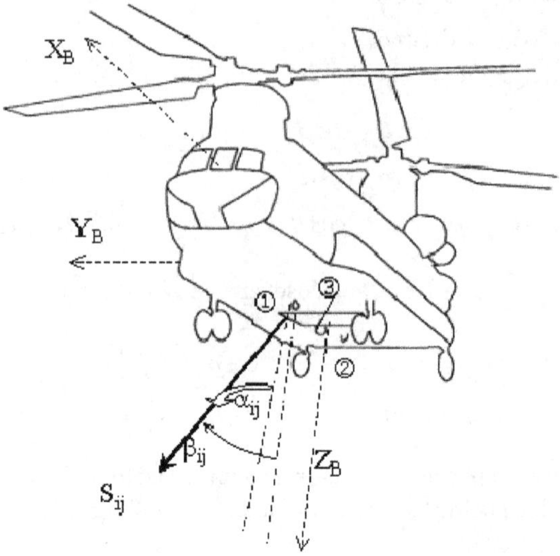

Fig. A6. Hooks and slings loads

7.2 Load model

The load is modelled as a 6-dof body, the helicopter slings being connected to the load as seen in Fig. A7.

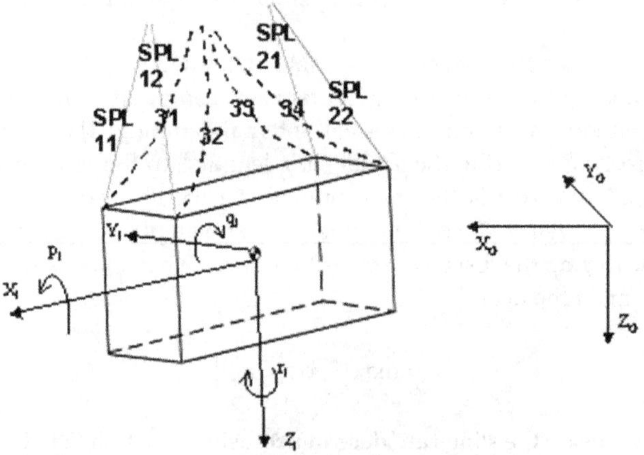

Fig. A7. Container load with local axes and sling numbering

The forces and moments acting on the load centre of gravity, besides load gravitational forces, consist of cable forces indexed with C and aerodynamic forces, which are determined using flat plate theory, indexed *Bl*. The summation of these forces and moments is given by equation (29).

$$
\begin{aligned}
X_l &= X_C + X_{Bl} \\
Y_l &= Y_C + Y_{Bl} \\
Z_l &= Z_C + Z_{Bl} \\
L_l &= L_C + L_{Bl} \\
M_l &= M_C + M_{Bl} \\
N_l &= N_C + N_{Bl}
\end{aligned}
\tag{29}
$$

The contributions of each cable and of the load aerodynamics are given in equations (30) and (31) respectively. Note that the cable component is valid only for a V-suspension.

$$
\begin{aligned}
X_C &= S_{ij} \cos\alpha_{lij} \sin\beta_{lij} \\
Y_C &= S_{ij} \sin\alpha_{lij} \sin\beta_{1i1} \\
Z_C &= -S_{ij} \cos\beta_{lij} \\
L_C &= S_{ij} \sin\alpha_{lij} \sin\beta_{lij} h_{cgl} - S_{i1} \cos\beta_{li1} w_{cgl} + S_{i2} \cos\beta_{li2} \left(w_l - w_{cgl}\right) \\
M_C &= -S_{ij} \cos\alpha_{lij} \sin\beta_{lij} h_{cgl} + S_{1j} \cos\beta_{l1j} l_{cgl} - S_{2j} \cos\beta_{l21} \left(l_l - l_{cgl}\right) \\
N_C &= +S_{1j} \sin\alpha_{l1j} \sin\beta_{l1j} l_{cgl} - S_{2j} \cos\alpha_{l2j} \sin\beta_{l2j} \left(l_l - l_{cgl}\right) \\
&-S_{i1} \cos\alpha_{li1} \sin\beta_{li1} w_{cgl} + S_{i2} \cos\alpha_{li2} \sin\beta_{li2} \left(w_l - w_{cgl}\right)
\end{aligned}
\tag{30}
$$

The magnitude of the sling forces is given by equation (27). The load aerodynamics are taken from the flat plate theory, using the empirical equations of reference (Ostroff, Downing and Rood, 1976), see equation (31).

$$
\begin{aligned}
X_{Bl} &= -C_{FEl} \tfrac{1}{2} \rho V_l^2 \\
Y_{Bl} &= -C_{Y\beta l} \tfrac{1}{2} \rho V_l^2 \sin\beta_{sl} \\
Z_{Bl} &= -C_{L\alpha l} \tfrac{1}{2} \rho V_l^2 \sin\alpha_{sl} \\
L_{Bl} &= -C_{L\beta l} \tfrac{1}{2} \rho V_l^2 \sin\beta_{sl} \left|\cos\beta_{sl}\right| \left(1 - \left|\sin\alpha_{sl}\right|\right) \\
M_{Bl} &= C_{M\alpha} \tfrac{1}{2} \rho V_l^2 \sin\alpha_{sl} \cos\alpha_{sl} \\
N_{Bl} &= -C_{N\beta} \tfrac{1}{2} \rho V_l^2 \sin\beta_{sl} \cos\beta_{sl} \left(0.94 \sin\alpha_{sl} + 0.342 \cos\alpha_{sl}\right)
\end{aligned}
\tag{31}
$$

8. Appendix B

Data for the helicopter and load used in the simulation model

Chinook dimensions (see Fig. B1)	
h_f	2.093 m
h_r	3.527 m
l_f	6.425 m
l_r	5.450 m
f_1	0 m
fs_1	2.28092 m
fs_2	1.78308 m
fs_3	0.19812 m
$hs_{1,2}$	1.309 m
hs_3	1.509 m

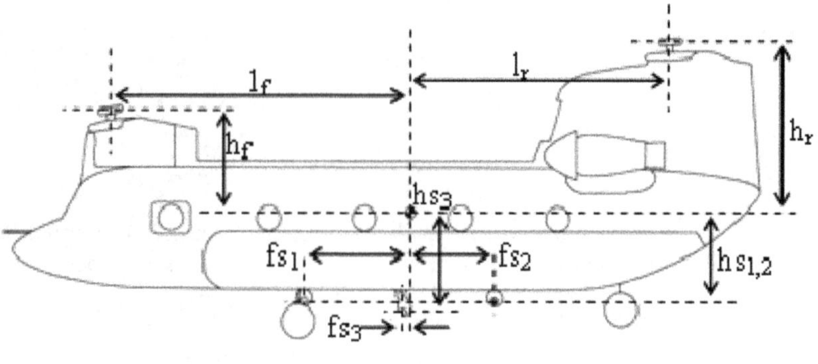

Fig. B1. Chinook dimensions

Rotor characteristics	
m_{bl}	161 kg
σ	0.08459
N	3
i_f	9 deg
i_r	4 deg
Ω	23.562 rad/s
γ	12.8
θ_{twist}	- 9.14 deg
R	9.144 m
c	0.81m
$C_{l\alpha}$	5.75 rad^{-1}
C_{do}	0.0098
C_{dt}	38.66

Chinook mass and moments of inertia	
m	14968.6 kg
I_x	50386.3 kg m^2
I_y	273536 kg m^2
I_z	257685 kg m^2
I_{xz}	19838.3 kg m^2
I_{bl}	3110.2 kg m^2

Chinook CH-47B aerodynamics	
C_{FE}	3.5 m^2
$C_{L\alpha\ fus}$	32.5 m^2
$C_{L\beta}$	6.57 m^3
$C_{M\alpha}$	142 m^3
$C_{N\beta}$	51.5 m^3
$C_{Y\beta}$	43.4 m^2

Blade control angle to stick displacement conversion factors	
Collective front rotor	0.734 deg/cm
Collective rear rotor	0.734 deg/cm
Longitudinal front rotor	0.242 deg/cm
Longitudinal rear rotor	-0.242 deg/cm
Lateral front rotor	0.752 deg/cm
Lateral rear rotor	-0.752 deg/cm
Pedal front rotor	1.25 deg/cm
Pedal rear rotor	- 1.25 deg/cm
Collective front/rear neutral	7.85 deg

Blade control system limits	deg	cm stick
Collective range front rotor	0 – 16 deg	0 – 21.8 cm
Collective range rear rotor	0 – 16 deg	
Longitudinal cyclic range front rotor	- 0.5 – 2 deg	-2 – 10.33 cm
Longitudinal cyclic range rear rotor	- 0.5 – 2.5 deg	
Lateral cyclic angle front rotor	- 10 – 10 deg	-13 – 13 cm
Lateral cyclic angle rear rotor	- 10 – 10 deg	

Container load mass and dimensions	
m_l	2000, 6000 and 10000 kg
l_l	6.058 m
w_l	2.438 m
h_l	2.591 m
$l_{CG\,l}$	(½ l_l) m
$w_{CG\,l}$	(½ w_l) m
h_{CGl}	(½ h_l) m

Load moments of inertia	
$I_{xx\,l}$	$\frac{1}{12} m_l \left(w_l^2 + h_l^2\right)$
$I_{yy\,l}$	$\frac{1}{12} m_l \left(h_l^2 + l_l^2\right)$
$I_{zz\,l}$	$\frac{1}{12} m_l \left(w_l^2 + l_l^2\right)$
$I_{xz\,l}$	0

Load moments of inertia	
$I_{xx\,l}$	$\frac{1}{12} m_l \left(w_l^2 + h_l^2\right)$
$I_{yy\,l}$	$\frac{1}{12} m_l \left(h_l^2 + l_l^2\right)$
$I_{zz\,l}$	$\frac{1}{12} m_l \left(w_l^2 + l_l^2\right)$
$I_{xz\,l}$	0

Sling aerodynamics	
k	$7.25 \cdot 10^5$ N/m
k_{red}	$1 \cdot 10^5$ m^2/rad
$l_{normal\,ij}$	4 m
l_{red}	5.2 m
c_{damp}	10 kg/s

9. Acknowledgment

The subject of this research was initiated as collaboration between Delft University of Technology, TNO Defence, Security and Safety and Royal Netherlands Air Force (RNLAF). Sincere thanks for all support given by the Chinook pilots to judge the results.

10. References

ADS-33 (2000) *Aeronautical Design Standard-33E-PRF, Performance Specification, Handling Qualities Requirements for Military Rotorcraft,* US Army AMCOM, Redstone, Alabama

Anon. (2002) *Technical Manual, Aviation Unit and Aviation Intermediate Maintenance Manual CH-47D Helicopter,* TM 55-1520-240-23-9, Headquarters, Department of the Army, Rev. 19 Sept. 2002, chapter 11, section IX, page 1012

Anon, (2004), *Boeing Integrated Defense Systems, ICH-47D-10: Flight Manual International CH-47D Helicopter,* The Boeing Company, 2004

Bisgaard, Morten, Bendtsen, Jan Dimon, la Cour-Habo, Anders (2006) Modelling of Generic Sling Load System, AIAA Modeling and Simulation Technologies Conference and Exhibit, 21-24 Aug. 2006, Keystone, Colorado, AIAA 2006-6816

Boeing Helicopters (2004), *Flight Manual International CH-47D Helicopter,* The Boeing Company, Philadelphia, Pennsylvania, Rev 4, 30 September 2004.

Chen, C., Lim, K.Y. and Seah, C.S.P.(1998) Modeling and Dynamic Analysis of Helicopter Underslung System", *AIAA Modeling and Simulation Technologies Conference and Exhibit,* 10-12 Aug. 1998, Boston, MA, AIAA-98-4358

Cicolani, Luigi S., Kanning, Gerd, Synnestvedt, Robert (1995) Simulation of the Dynamics of Helicopter Slung Load Systems. *J. of the American Helicopter Society,* Vol. 40, No. 4, October 1995, pp. 44-61

Cliff, E.M. and Bailey (1975) D.B., Dynamic Stability of a Translating Vehicle with a Simple Sling Load, *J. of Aircraft,* Vol. 12, No. 10, Oct. 1975

Dukes, Theodor, A. (1973) Maneuvring Heavy Sling Loads Near Hover Part II: Some Elementary Maneuvres, *J. of the American Helicopter Society*, Vol. 18, No. 3, July 1973, pp. 3-13

Dukes, Theodor, A. (1973) Maneuvring Heavy Sling Loads Near Hover Part II: Some Elementary Maneuvres, *J. of the American Helicopter Society*, Vol. 18, No. 3, July1 973, pp. 3-13

Einthoven P., et al. (2006) Development of Control Laws for the Chinook Digital AFCS program, *62nd American Helicopter Society Annual Forum*, May 9-11, 2006, Phoenix, Arizona

Fusato, Dario, Guglieri, Giorgio, Celi, Roberto (2001) Flight Dynamics of an Articulated Rotor Helicopter with an External Load, *J. of the American Helicopter Society*, Vol. 46, No. 1, January 2001, pp. 3-12

Gabel, Richard, Wilson, Gregory, J. (1968). Test Approaches to External Sling Load Instabilities, *J. of American Helicopter Society*, Vol. 13, No.3, July 1968, pp. 44-54

Kendrick, Stephen, A., Walker, Daniel, J. (2006) The Modelling, Simulation and Control of Helicopters Operating with External Loads, *62nd Annual Forum of the American Helicopter Society*, May 9-11, 2006, Phoenix, AZ

Lucassen, L.R., and Stenk F.J.(1965) Dynamic Stability Analysis of a Hovering Helicopter with a Slung Load", *J. of the American Helicopter Society*, 10(2), pp.. 1965

Nagabhushan, B.L. (1985). Low-Speed Stability Characteristics of a Helicopter with a Sling Load", *VERTICA*, Vol. 9, No.4, 1985, pp. 345-361

Ostroff A.J., Downing, D.R. And Rood, W.J. (1976) A technique using a non-linear helicopter model for determining trims and derivatives, NASA TN D-8159, Langley Research Center Hampton, Va, May 1976

Pavel, M.D.(2007) Mathematical Modeling of Tandem Helicopters with External Sling Loads for Piloted Simulation, AIAA Modeling and Simulation Technologies Conference and Exhibit, Hilton Head, South Carolina, IAA-2007-6617

Pavel, M.D. (2010) Investigation of Chinook helicopter operations with an external slung load after cable failure. *The Aeronautical Journal*, 114(1156), pp. 345-365

Feaster, L., Poli, C. & Kirchhoff, R. (1977) Dynamics of a Slung Load, *J. of Aircraft*, Vol. 14, No. 2, pp. 115-121

Prabhakar, A. (1978) Stability of a Helicopter Carrying an Underslung Load, *VERTICA*, Vol.2, No.2, 1978, pp. 121-143

Prouty, R.W. (2001) *Longitudinal cyclic control for tandem-rotor helicopters*, VERTIFLITE magazine, vol. 47, no. 3. 2001

Reijm, Rogier, Pavel, Marilena D&, Bart, Jacob (2006), Effects of Automatic Flight Control System on Chinook Underslung Failures, *32nd European Rotorcraft Forum*, Maastricht, The Netherlands, September, 12-14, 2006

Sheldon, D.F. (1977) An Appreciation of the Dynamic Problems Associated with the External Transportation of Loads from a Helicopter-State of the Art, *VERTICA*, Vol. 1, 1977

Stuckey, R.A. (2002) *Mathematical Modelling of helicopter Slung-load systems*, DSTO Aeronautical and Maritime Research Laboratory, Victoria Australia, 2002

Tyson, Peter, et. al. (1999). Simulation Prediction and Flight Validation of UH-60A Black Hawk Slung load Characteristics, *55th American Helicopter Society Annual Forum*, May 25-27, 1999, Montreal, Canada

Van der Kamp, Reinier., et. al. (2005). Investigation on the Chinook operations with an external sling load after cable failure", *31st European Rotorcraft Annual Forum,* Florence, Italy, September 13-15, 2005

7

Tool-Based Design and Evaluation of Resilient Flight Control Systems

Hafid Smaili[1], Jan Breeman[2] and Thomas Lombaerts[3]
National Aerospace Laboratory NLR/Delft University of Technology
[1]NLR Cockpit & Flight Operations Department, Amsterdam,
[2]NLR Avionics & Systems Department, Amsterdam,
[3]DLR Institute of Robotics and Mechatronics, Wessling,
[1,2]The Netherlands
[3]Germany

1. Introduction

A large transport aircraft simulation benchmark (REconfigurable COntrol for Vehicle Emergency Return RECOVER) has been developed within the European GARTEUR Flight Mechanics Action Group 16 (FM-AG(16)) on Fault Tolerant Control (2004-2008) for the integrated evaluation of fault detection, identification (FDI) and reconfigurable flight control systems. The benchmark includes a suitable set of assessment criteria and failure cases, based on reconstructed accident scenarios, to assess the potential of new adaptive control strategies to improve aircraft survivability. The application of reconstruction and modeling techniques, using accident flight data for validation, has resulted in high fidelity non-linear aircraft and fault models to evaluate new Fault Tolerant Flight Control (FTFC) concepts and their real-time performance to accommodate in-flight failures (Edwards et al., 2010).

This chapter will give an overview of advanced flight control developments and pilot training related initiatives to reduce the amount of in-flight loss-of-control (LOC-I) accidents. The GARTEUR RECOVER benchmark, validated with accident flight data and used during the GARTEUR FM-AG(16) program, will be described. The modular features of the benchmark will be outlined that address the need for tool-based design of modern resilient flight control systems that mitigate potentially catastrophic (mechanical) failures and aircraft upsets.

2. Program overview

Fault tolerant flight control (FTFC) enables improved survivability and recovery from adverse flight conditions induced by faults, damage and associated upsets. This can be achieved by "intelligent" utilisation of the control authority of the remaining control effectors in all axes consisting of the control surfaces and engines or a combination of both. In this technique, control strategies are applied to restore stability and maneuverability of the vehicle for continued safe operation and a survivable landing.

From 2004-2008, a research group on Fault Tolerant Control, comprising a collaboration of thirteen European partners from industry, universities and research institutions, was established within the framework of the Group for Aeronautical Research and Technology in Europe (GARTEUR) co-operation program (Table 1). The aim of the research group, Flight Mechanics Action Group FM-AG(16), is to demonstrate the capability and potential of innovative reconfigurable flight control algorithms to improve aircraft survivability. The group facilitated the proliferation of new developments in fault tolerant control design within the European aerospace research and academic community towards practical and real-time operational applications. This addresses the need to improve the resilience and safety of future aircraft and aiding the pilot to recover from adverse conditions induced by (multiple) system failures, damage and (atmospheric) upsets that would otherwise be potentially catastrophic. Up till now, faults or damage on board of aircraft have been accommodated by hardware design using duplex, triplex or even quadruplex redundancy of critical components. The approach of the GARTEUR research focussed on providing redundancy by means of new adaptive control law design methods to accommodate (unanticipated) faults and/or damage that dramatically change the configuration of the aircraft. These methods take into account a novel combination of robustness, reconfiguration and (real-time) adaptation of the control laws (Edwards et al., 2010; Lombaerts et al., 2009).

GARTEUR
QinetiQ, Bedford, United Kingdom
Airbus, Toulouse, France
National Aerospace Laboratory (NLR), Amsterdam, The Netherlands
Deutsches Zentrum für Luft- und Raumfahrt (DLR), Braunschweig and Oberpfaffenhofen, Germany
Defence Science and Technology Laboratory (DSTL), Bedford, United Kingdom
Centro Italiano Ricerche Aerospaziali (CIRA), Capua, Italy
Delft University of Technology, Delft, The Netherlands
Cambridge University, Cambridge, United Kingdom
Aalborg University, Esbjerg, Denmark
University of Lille, Lille, France
University of Hull, Hull, United Kingdom
University of Bordeaux, Bordeaux, France
University of Leicester, Leicester, United Kingdom

Table 1. GARTEUR Flight Mechanics Action Group 16 (FM-AG(16)) Fault Tolerant Control consortium

The group addressed the need for high-fidelity nonlinear simulation models, relying on accurate failure modelling, to improve the prediction of reconfigurable system performance in degraded modes. As part of this research, a simulation benchmark was developed, based on the Boeing 747-100/200 large transport aircraft, for the assessment of fault tolerant flight control methods. The test scenarios that are an integral part of the benchmark were selected to provide challenging assessment criteria to evaluate the effectiveness and potential of the FTFC methods being investigated (Lombaerts et al., 2006) The simulation model of the benchmark was earlier applied in an investigation of the 1992 Amsterdam Bijlmermeer airplane accident (Flight 1862) (Netherlands Aviation Safety Board, 1994; Smaili, 1997, 2000) and has been validated against data from the Digital Flight Data Recorder (DFDR) of the accident flight.

The potential of the developed fault tolerant flight control methods to improve aircraft survivability, for both manual and automatic flight, has been demonstrated in 2008 during a piloted assessment in the SIMONA research flight simulator of the Delft University of Technology (Edwards et al., 2010; Stroosma et al., 2009).

3. Fault tolerant flight control systems

An increasing number of measures are currently being taken by the international aviation community to prevent Loss Of Control In-Flight (LOC-I) accidents due to failures, damage and upsets for which the pilot was not able to recover successfully despite the available performance and control capabilities. Recent airliner accident and incident statistics (Civil Aviation Authority of the Netherlands (CAA-NL), 2007) show that about 16% of the accidents between the 1993 and 2007 period can be attributed to LOC-I, caused by a piloting mistake, technical malfunction or unusual upsets due to external (atmospheric) disturbances. However, worldwide civil aviation safety statistics indicate that today 'in-flight loss of control' has become the main cause of aircraft accidents (followed by 'controlled flight into terrain' (CFIT)). Data examined by the international aviation community shows that, in contrast to CFIT, the share of LOC-I occurrences is not significantly decreasing. The actions taken by the aviation community to lower the number of LOC-I occurrences not only include improvements in procedures training and human factors, but also finding measures to better mitigate system failures and increase aircraft survivability in case of an accident or degraded flight conditions.

Reconfigurable flight control, or "intelligent flight control", is aimed to prevent aircraft loss due to multiple failures when the aircraft is still flyable given the available control power. Motivated by several aircraft accidents at the end of the 1970's, in particular the crash of an American Airlines DC-10 (Flight 191) at Chicago in 1979, research on "self-repairing", or reconfigurable fault tolerant flight control (RFTFC), was initiated to accommodate in-flight failures. Reconfigurable control aims to utilise all remaining control effectors on the aircraft after a (unanticipated) mechanical or structural failure to recover the performance of the original system by automatic redesign of the flight control system. The first objective of reconfiguration is to guarantee system stability while the original performance is reconstructed as much as possible. Due to limitations of the control allocation scheme caused by, for instance, actuator position and rate limits, the system performance of the unfailed aircraft may not be fully achieved. In this case, the failed aircraft would be flown in a degraded mode but with sufficiently acceptable handling qualities for a successful recovery. Reconfigurable flight control systems have been successfully flight tested and evaluated in manned simulations (Burcham & Fullerton, 2004; Corder, 2004; Ganguli et al., 2005; The Boeing Company, 1999; Wright Laboratory, 1991).

Adaptive or reconfigurable flight control strategies might have prevented the loss of two Boeing 737s due to a rudder actuator hardover and of a Boeing 767 due to inadvertent asymmetric thrust reverser deployment. The 1989 Sioux City DC-10 incident is an example of the crew performing their own reconfiguration using asymmetric thrust from the two remaining engines to maintain limited control in the presence of total hydraulic system failure. Following the Sioux City incident in 1989, during which the engines were used as only remaining control effectors after loss of all hydraulics, a program was initiated at the NASA Dryden Flight Research Center on Propulsion Controlled Aircraft (PCA) (Burcham 2004). The system aims to provide a safe landing capability using only augmented engine thrust for flight control. Throughout the 1990's, the system has been successfully tested on several aircraft, including both commercial and military (Figure 1).

Fig. 1. McDonnell Douglas MD-11 landing at Dryden Flight Research Center equipped with a computer-assisted engine control landing system developed by NASA (NASA Dryden photo collection)

The crash of a Boeing 747 freighter in 1992 near Amsterdam, the Netherlands, following the separation of the two right-wing engines, was potentially survivable given adequate knowledge about the remaining aerodynamic capabilities of the damaged aircraft (Smaili, 1997, 2000). New kinds of threats within the aviation community have recently been introduced by deliberate hostile attacks on both commercial and military aircraft. For instance, a surface-to-air missile (SAM) attack has recently been demonstrated to be survivable by the crew of an Airbus A300B4 freighter performing a successful emergency landing at Baghdad International Airport after suffering from complete hydraulic system failures and severe structural wing damage (Figure 2).

Fig. 2. Emergency landing sequence using engines only and left wing structural damage due to surface-to-air missile impact, DHL A300B4-203F, Baghdad, 2003

Apart from system failures and hostile actions against commercial and military aircraft, recent incident cases also show the destructive impact of hazardous atmospheric weather conditions on the structural integrity of the aircraft. In some cases, clear air turbulence has resulted in aircraft incurring substantial structural damage and loss of engines due to clear air turbulence (CAT).

A number of new fault detection and isolation methods have been proposed in the literature (Patton, 1997; Zhang & Jiang, 2003, Zhang, 2005) together with methods for reconfiguring flight control systems. To assess these new methods for aerospace applications, they need to be integrated and applied to realistic operational scenarios that include representative levels of non-linearity, noise and disturbance. This will then allow the benefits of these new flight control technologies to be evaluated in terms of functionality and performance.

Studies of airliner LOC-I accidents (Edwards, 2010; Smaili, 1997, 2000) show that better situational awareness or guidance would have recovered the impaired aircraft and improved survivability if unconventional control strategies were used. In some of the cases studied, the crew was able to adapt to the unknown degraded flying qualities by applying control strategies (e.g. using the engines effectors to achieve stability and control augmentation) that are not part of any standard airline training curriculum.

The results of a LOC-I study concerning the 1992 Amsterdam accident case (Smaili, 1997, 2000), in which a detailed reconstruction and simulation of the accident flight was conducted based on the recovered Digital Flight Data Recorder (DFDR), formed the basis for realistic and validated aircraft accident scenarios as part of the GARTEUR FM-AG(16) aircraft simulation benchmark. The study resulted in high fidelity non-linear fault models for a civil large transport aircraft that addresses the need to improve the prediction of reconfigurable system performance in degraded modes.

4. Flight 1862 aircraft accident case

On October 4th, 1992, a Boeing 747-200F freighter aircraft, Flight 1862 (Figure 3), went down near Amsterdam Schiphol Airport after the separation of both right-wing engines. In an attempt to return to the airport for an emergency landing, the aircraft flew several right-hand circuits in order to lose altitude and to line up with the runway as intended by the crew. During the second line-up, the crew lost control of the aircraft. As a result, the aircraft crashed, 13 km east of the airport, into an eleven-floor apartment building in the Bijlmermeer, a suburb of Amsterdam. Results of the accident investigation, conducted by several organisations including the Netherlands Accident Investigation Bureau and the aircraft manufacturer, were hampered by the fact that the actual extent of the structural damage to the right-wing, due to the loss of both engines, was unknown. The analysis from this investigation concluded that given the performance and controllability of the aircraft after the separation of the engines, a successful landing was highly improbable (Smaili, 1997, 2000).

In 1997, the division of Control and Simulation of the Faculty of Aerospace Engineering of the Delft University of Technology (DUT), in collaboration with the Netherlands National Aerospace Laboratory NLR, performed an independent analysis of the accident (Smaili, 1997, 2000). In contrast to the analysis performed by the Netherlands Accident Investigation Bureau, the parameters of the DFDR were reconstructed using comprehensive modelling, simulation and visualisation techniques. In this alternative approach, the DFDR pilot control inputs were applied to detailed flight control and aerodynamic models of the accident aircraft. The purpose of the analysis was to acquire an estimate of the actual flying capabilities of the aircraft and to study alternative (unconventional) pilot control strategies for a successful recovery. The application of this technique resulted in a simulation model of the impaired aircraft that could reasonably predict the performance, controllability effects and control surface deflections as observed on the DFDR. The analysis of the reconstructed model of the aircraft, as used for the GARTEUR FM-AG(16) benchmark, indicated that from a flight mechanics point of view, the Flight 1862 accident aircraft was recoverable if unconventional control strategies were used (Smaili, 1997, 2000).

Fig. 3. Cargo accident aircraft prior to takeoff at Amsterdam Schiphol Airport (left). Reconstructed loss of control based on flight data following separation of the right-wing engines (right), EL AL Flight 1862, B747-200F, Amsterdam, 1992 (copyright Werner Fischdick, NLR)

4.1 Aircraft damage configuration

The Flight 1862 damage configuration to both the aircraft's structure and onboard systems, after the separation of both right-wing engines, is illustrated in Figure 4. An analysis of the engine separation dynamics concluded that the sequence was initiated by the detachment of the right inboard engine and pylon (engine no. 3) from the main wing due to a combination of structural overload and metal fatigue in the pylon-wing joint. Following detachment, the right inboard engine struck the right outboard engine (engine no. 4) in its trajectory also rupturing the right-wing leading edge up to the front spar. The associated loss of hydraulic systems resulted in limited control capabilities due to unavailable control surfaces aggravated by aerodynamic disturbances caused by the right-wing structural damage.

The crew of Flight 1862 was confronted with a flight condition that was very different from what they expected based on training. The Flight 1862 failure mode configuration resulted in degraded flying qualities and performance that required adaptive and unconventional (untrained) control strategies. Additionally, the failure mode configuration caused an unknown degradation of the nominal flight envelope of the aircraft in terms of minimum control speed and manoeuvrability. For the heavy aircraft configuration at a relative low speed of around 260 knots IAS, the DFDR indicated that flight control was almost lost requiring full rudder pedal, 60 to 70 percent maximum control wheel deflection and a high thrust setting on the remaining engines.

Flight 1862 Failure Mode Configuration
Aircraft Systems
Hydraulic systems #3 and #4 lost
Engine #3 and #4 thrust loss
Lower rudder lag
Mass Properties
Engine #3 and #4 weight loss, 4,014 kg each
Pylon #3 and #4 weight loss, ± 1,000 kg each
Lateral center of gravity displacement
Total weight loss: 10,0028 kg
Aerodynamics
Lift loss due to wing damage (ΔL_{wing})
Rolling moment due to wing damage ($\Delta \bar{L}_{wing}$)
Drag due to wing damage (ΔD_{wing})
Yawing moment due to wing damage ($\Delta \bar{N}_{wing}$)
Pitching moment due to wing damage (ΔM_{wing})
Right inboard aileron and spoiler #10 and #11 aerodynamic efficiency loss

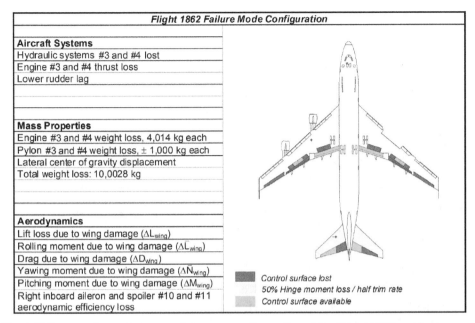

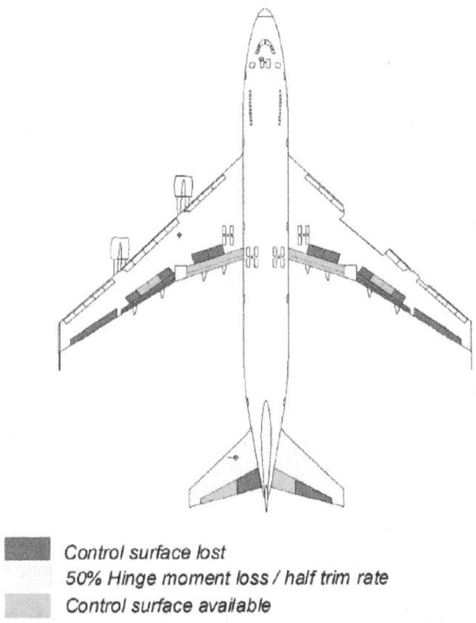

Fig. 4. Failure modes and structural damage configuration of the Flight 1862 accident aircraft suffering right-wing engine separation, partial loss of hydraulics and change in aerodynamics

4.2 Aircraft survivability assessment

Figure 5 presents the performance capabilities of the Flight 1862 accident aircraft after separation of both right-wing engines, reconstructed via the methods described in (Edwards et al., 2010; Smaili, 1997, 2000), as a function of thrust and aircraft weight. The reconstructed model indicates that in these conditions and at heavy weight (700,000 lbs / 317,460 kg), level flight capability was available between Maximum Continuous Thrust (MCT) and Take-Off/Go Around thrust (TOGA). At or above approximately TOGA thrust, the aircraft had limited climb capability. Analysis shows that adequate control capabilities remained available to achieve the estimated performance capabilities. Figure 5 indicates a significant improvement in available performance and controllability at a lower weight if more fuel had been jettisoned.

Simulation analysis of the accident flight using the reconstructed model (Edwards et al., 2010; Smaili, 1997, 2000) predicts sufficient performance and controllability, after the separation of the engines, to fly a low-drag approach profile at a 3.5 degrees glide slope angle for a high-speed landing or ditch at 200/210 KIAS and at a lower weight. Note again that this lower weight could have been obtained by jettisoning more fuel. The lower thrust requirement for this approach profile results in a significant improvement in lateral control margins that are adequate to compensate for additional thrust variations. The above predictions have been confirmed during the piloted simulator campaign later in the GARTEUR FM-AG(16) program.

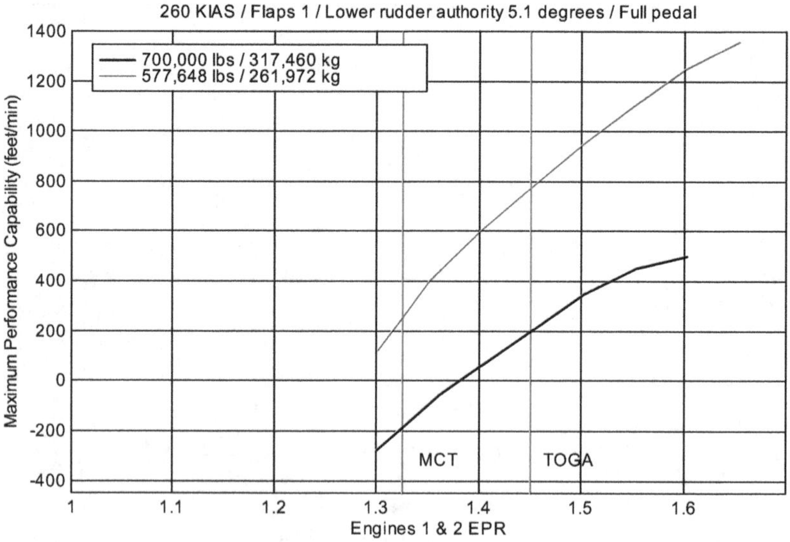

Fig. 5. Flight 1862: Effect of engine thrust and weight on maximum climb performance for straight flight at 260 KIAS

5. GARTEUR RECOVER benchmark

For the assessment of novel fault tolerant flight control techniques, the GARTEUR FM-AG (16) research group developed a simulation benchmark, based on the reconstructed Flight 1862 aircraft model (REconfigurable COntrol for Vehicle Emergency Return RECOVER). The benchmark simulation environment is based on the Delft University Aircraft Simulation and Analysis Tool DASMAT. The DASMAT tool was further enhanced with a full nonlinear simulation of the Boeing 747-100/200 aircraft (Flightlab747 / FTLAB747), including flight control system architecture, for the Flight 1862 accident study as conducted by Delft University. The simulation environment was subsequently utilised and further enhanced as a realistic tool for evaluation of fault detection and fault tolerant control schemes within other research programmes (Marcos & Balas, 2001). Reference (Edwards et al., 2010) provides details on the model reconstruction and validation based on the Flight 1862 accident data and simulation model implementations. For the application of the benchmark model, reference (Edwards et al., 2010) also provides a description regarding the benchmark model architecture, mathematical models and user examples.

The GARTEUR RECOVER benchmark has been developed as a Matlab®/Simulink® platform for the design and integrated (real-time) evaluation of new fault tolerant control techniques (Figure 6, 7 and 8). The benchmark consists of a set of high fidelity simulation and flight control design tools, including aircraft fault scenarios. For a representative simulation of damaged aircraft handling qualities and performances, the benchmark aircraft model has been validated against data from the Digital Flight Data Recorder (DFDR) of the EL AL Flight 1862 Boeing 747-200 accident aircraft that crashed near Amsterdam in 1992 caused by the separation of its right-wing engines.

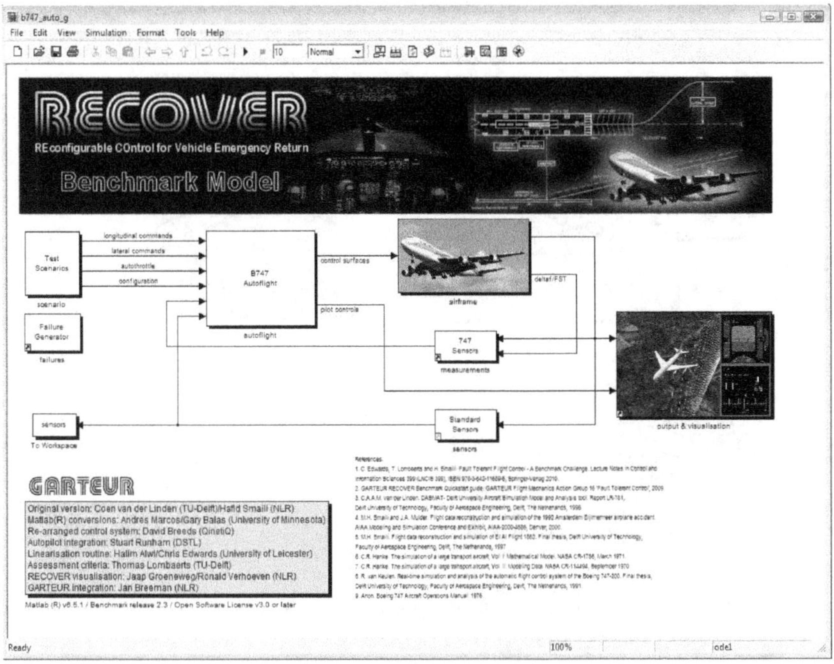

Fig. 6. GARTEUR RECOVER Benchmark main model components for closed-loop simulations

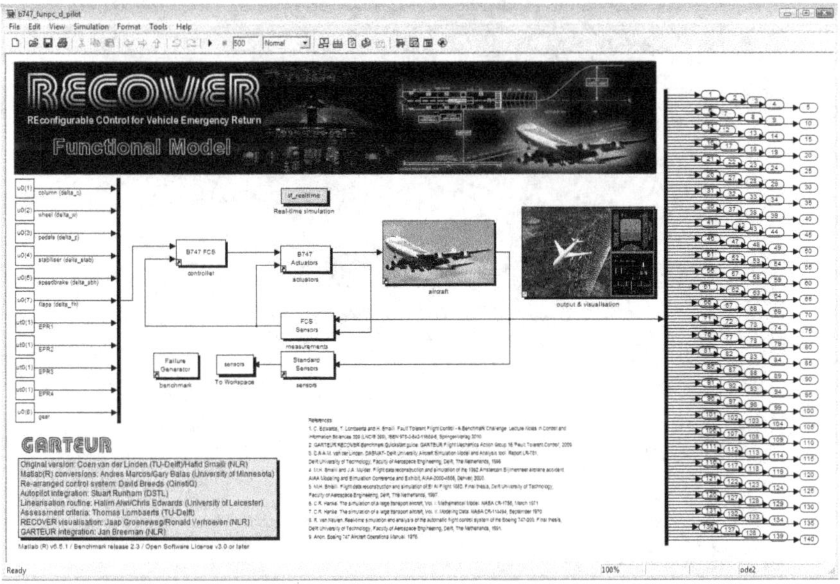

Fig. 7. GARTEUR RECOVER Benchmark functional model for open-loop nonlinear off-line (interactive) simulations

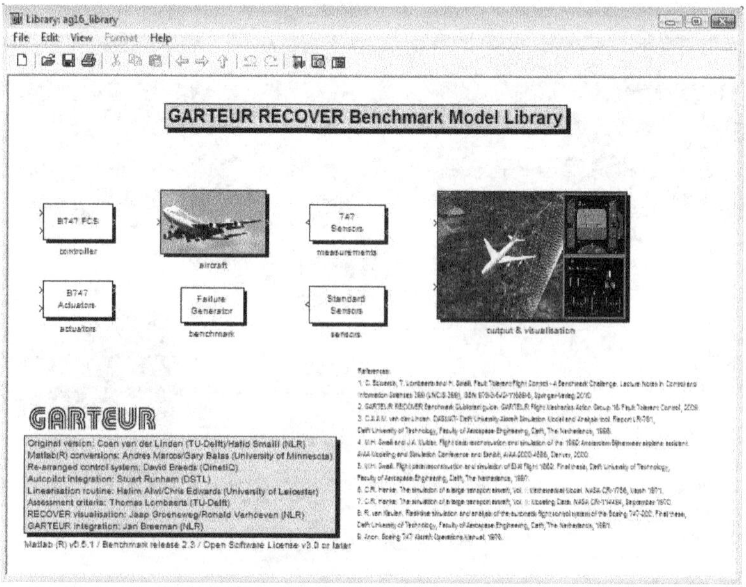

Fig. 8. GARTEUR RECOVER Benchmark component library

Fig. 9. GARTEUR RECOVER Benchmark graphical user interface for the selection of simulation and analysis tools

The GARTEUR RECOVER benchmark software package is equipped with several simulation and analysis tools, all centered around a generic non-linear aircraft model for six degrees-of-freedom non-linear aircraft simulations. For high performance computation and visualisation capabilities, the package has been integrated as a toolbox in the computing environment Matlab®/Simulink®. The benchmark is operated via a Matlab® graphical user interface (Figure 9) from which the different benchmark tools may be selected. The user options in the main menu are divided into three main sections allowing to initialise the benchmark, run the simulations and select the analysis tools. The tools of the GARTEUR RECOVER benchmark include trimming and linearisation for (fault tolerant) flight control law design, nonlinear off-line (interactive) simulations, simulation data analysis and flight trajectory and pilot interface visualisations (Figure 10). The modularity of the benchmark makes it customisable to address research goals in terms of aircraft type, flight control system configuration, failure scenarios and flight control law assessment criteria.

The test scenarios that are an integral part of the GARTEUR RECOVER benchmark were selected to provide challenging (operational) assessment criteria, as specifications for reconfigurable control, to evaluate the effectiveness and potential of the FTFC methods being investigated in the GARTEUR program. Validated against data from the DFDR, the benchmark provides accurate aerodynamic and flight control failure models, realistic scenarios and assessment criteria for a civil large transport aircraft with fault conditions ranging in severity from major to catastrophic.

The geometry of the GARTEUR RECOVER benchmark flight scenario (Figure 11) is roughly modelled after the Flight 1862 accident profile. The scenario consists of a number of phases. First, it starts with a short section of normal flight after which a fault occurs, which is in turn followed by a recovery phase. If this recovery is successful, the aircraft should again be in a stable flight condition, although not necessarily at the original altitude and heading. After recovery, an optional identification phase is introduced during which the flying capabilities of the aircraft can be assessed. This allows for a complete parameter identification of the model for the damaged aircraft as well as the identification of the safe flight envelope. The knowledge gained during this identification phase can be used by the controller to improve the chances of a safe landing. In principle, the flight control system is now reconfigured to allow safe flight. The performance of the reconfigured aircraft is subsequently assessed in a series of five flight phases. These consist of straight and level flight, a right-hand turn to a course intercepting the localizer, localizer intercept, glideslope intercept and the final approach. During the final approach phase, the aircraft is subjected to a sudden lateral displacement just before the threshold, which simulates the effect of a low altitude windshear. The landing itself is not part of the benchmark, because a realistic aerodynamic model of the damaged aircraft in ground effect is not available. However, it is believed that if the aircraft is brought to the threshold in a stable condition, the pilot will certainly be able to take care of the final flare and landing.

The GARTEUR RECOVER benchmark simulation model, as applied within this research program, is available via the website of the project after registration (www.faulttolerantcontrol.nl).

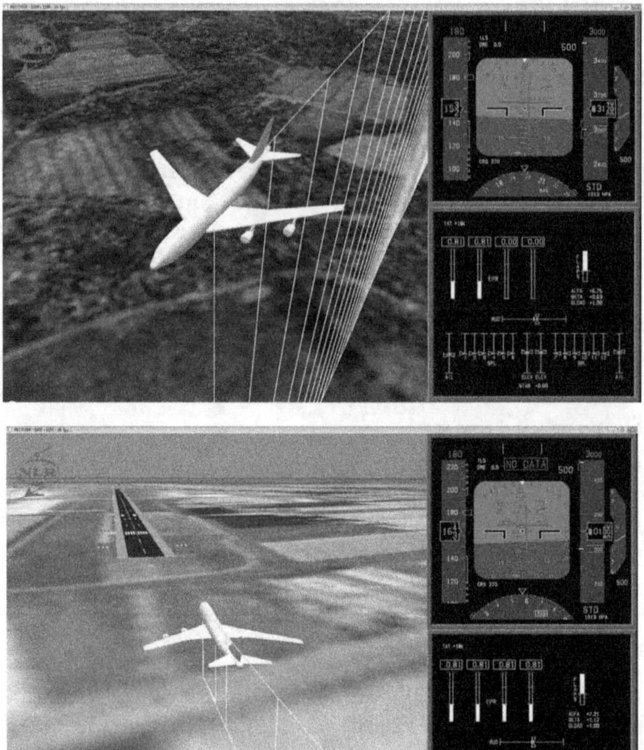

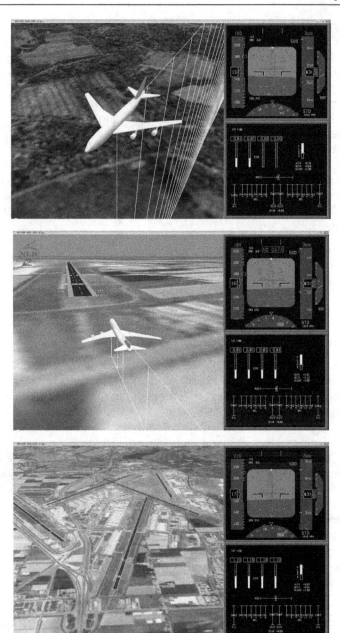

Fig. 10. GARTEUR RECOVER Benchmark high resolution aircraft visualisation tool for interactive (real-time) simulation and validation of new fault tolerant flight control algorithms. Tool features include pilot interface displays, environment modeling, aircraft flight path animation and detailed renditions of Amsterdam Schiphol airport as part of the benchmark approach and landing scenario

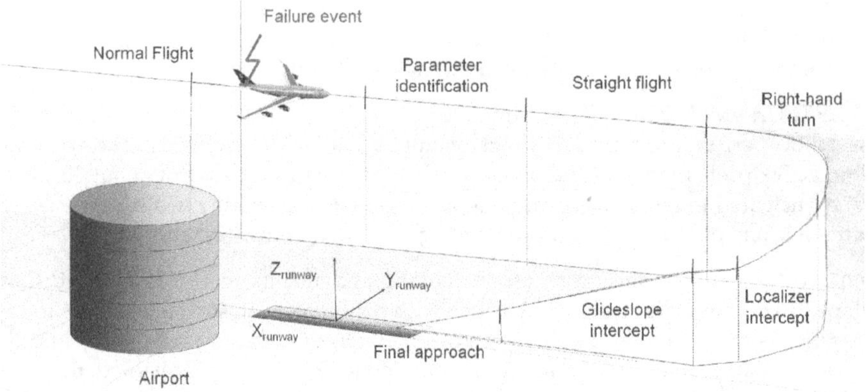

Fig. 11. GARTEUR RECOVER Benchmark flight scenario for qualification of fault tolerant flight control systems for safe landing of a damaged large transport aircraft (Edwards et al., 2010; Lombaerts et al., 2006)

6. Flight simulator integration and piloted assessment

The developed fault tolerant flight control schemes in this project have been evaluated in a piloted simulator assessment using a real-time integration of the GARTEUR RECOVER benchmark model, including reconstructed accident scenarios (Edwards et al., 2010; Stroosma et al., 2009). The evaluation was conducted in the SIMONA Research Simulation (SRS) facility, a full 6 degrees of freedom motion research simulator, of the Delft University of Technology (Figure 12).

Fig. 12. Evaluation of GARTEUR FM-AG(16) FTFC techniques in the Delft University SIMONA Research Simulator based on reconstructed accident scenarios (Left: Boeing 747 cockpit configuration. Right: visual system dome)

Several validation steps were performed to assure the benchmark model was implemented correctly. This included proof-of-match validation and piloted checkout of the baseline aircraft, control feel system and Flight 1862 controllability and performance characteristics. To accurately replicate the operational conditions of the reconstructed Flight 1862 accident aircraft in the simulator, the experiment scenario was aimed at a landing on runway 27 of Amsterdam Schiphol airport. The SIMONA airport scenery was representative of Amsterdam Schiphol airport and its surroundings for flight under visual flight rules (VFR).

The GARTEUR FM-AG(16) piloted simulator campaign provided a unique opportunity to assess pilot performance under flight validated accident scenarios and operational conditions. Six professional airline pilots, with an average experience of about 15.000 flight hours, participated in the piloted simulations. Five pilots were type rated for the Boeing 747 aircraft while one pilot was rated for the Boeing 767 and Airbus A330 aircraft.

In general, the results show, for both automatic and manual controlled flight, that the developed FTFC strategies were able to cope with potentially catastrophic failures in case of flight critical system failures or if the aircraft configuration has changed dramatically. In most cases, apart from any slight failure transients, the pilots commented that aircraft behaviour felt conventional after control reconfiguration following a failure, while the control algorithms were successful in recovering the ability to control the damaged aircraft. Manual controlled flight under fault reconfiguration was assessed for both a runaway of the rudder to the blow-down limit and a separation of both right-wing engines (Figure 13). Part of the FTFC strategies that were evaluated in the piloted simulation campaign consisted of a combination of real-time aerodynamic model identification and adaptive nonlinear dynamic inversion for control allocation and reconfiguration (Edwards et al., 2010; Lombaerts et al., 2009). The simulation results have shown that the handling qualities of the reconfigured damaged aircraft with a fault tolerant control system degrade less, indicating improved task performance. For both the Flight 1862 and rudder hardover case, as part of the scenarios surveyed in this research program, the pilots demonstrated the ability to fly the damaged aircraft, following control reconfiguration, back to the airport and conduct a survivable approach and landing (Edwards et al., 2010).

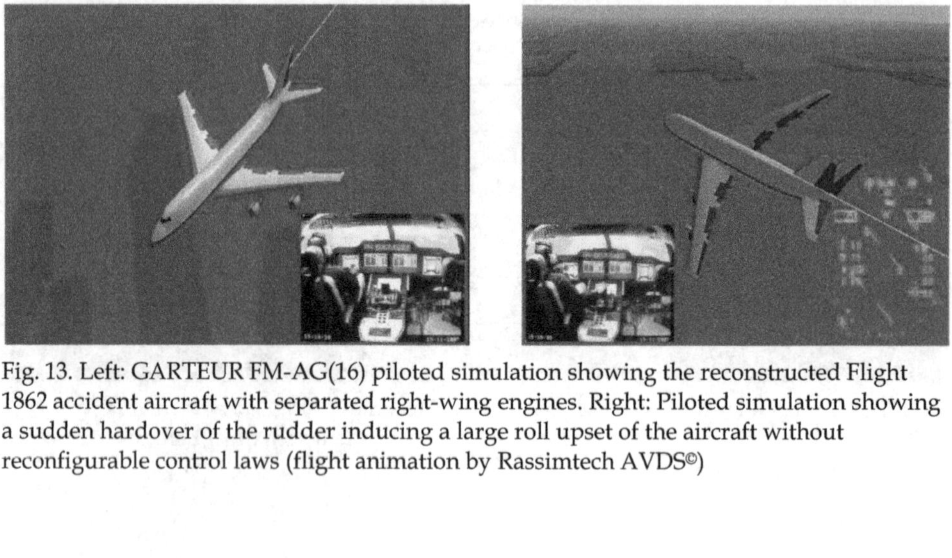

Fig. 13. Left: GARTEUR FM-AG(16) piloted simulation showing the reconstructed Flight 1862 accident aircraft with separated right-wing engines. Right: Piloted simulation showing a sudden hardover of the rudder inducing a large roll upset of the aircraft without reconfigurable control laws (flight animation by Rassimtech AVDS©)

7. Developments in aircraft loss-of-control prevention

The NASA Aviation Safety Program (NASA, 2011), which is a partnership between NASA and the Federal Aviation Administration (FAA), the Department of Defense (DoD) and the aviation industry, aims to further reduce the observed worldwide trends in aviation accidents by means of new loss-of-control prevention, mitigation and recovery techniques. These techniques, currently being investigated by the AvSP program apart from other measures, should assure to meet the demands of the transition to the Next Generation Air Transportation System (NextGen).

Future requirements from a flight deck system safety point of view include a more integrated design of information systems available to the pilot including displays and interactions, flight decision support systems (e.g. advisories during adverse and/or upset conditions including automatic recovery) and the allocation of functions between the pilot and automatic systems during nominal and degraded flight conditions. This new "intelligent" flight deck should be able to sense onboard (flight control) system and environmental-induced hazards in real-time and provide the necessary and timely actions to prevent or recover from any adverse condition (Figure 14).

Part of the technology strategies of the AvSP program include methods for improvements of vehicle system health-monitoring and survivability rate through "self-repairing" mechanisms in case of system failures. Within the AvSP Integrated Resilient Aircraft Control (IRAC) program (NASA, 2011), multidisciplinary integrated aircraft control design tools and techniques are investigated and developed to allow safe aircraft operation in the event of flight into adverse conditions (e.g. loss-of-control or upsets due to onboard control system failures, environmental factors or aerodynamic degradation caused by damage or icing). Adaptive flight control, as discussed in this chapter and investigated by the GARTEUR program, is provided within the IRAC program as a design option (in support of pilot training) to mitigate in-flight loss-of-control by enhancing the stability and maneuverability margins of the (damaged) aircraft for a safe and survivable landing. Additional applications of adaptive flight control might include the prevention or recovery of aircraft departures following inadvertent stall or unusual attitudes. These developments require accurate modeling of the dynamics involved in loss-of-control caused by failures or (post-departure) upset conditions in terms of system behaviour and aerodynamic characteristics. This requirement will allow representative simulation of dynamic flight conditions, based on wind tunnel data in combination with computational fluid dynamic (CFD) techniques (Figure 14), for adaptive control law design and evaluation.

Within the Active Management of Aircraft System Failures (AMASF) project, as part of the AvSP program, several issues in the area of FTFC technology have been addressed. These include detection and identification of failures and icing, pilot cueing strategies to cope with failures and icing and control reconfiguration strategies to prevent extreme flight conditions following a failure of the aircraft. In this context, a piloted simulation was conducted early 2005 of a Control Upset Prevention and Recovery System (CUPRSys). Despite few limitations, CUPRSys provided promising fault detection, isolation and reconfiguration capabilities (Ganguli, et al., 2005).

Fig. 14. Left: Future integrated "intelligent" flight deck for safe and efficient operation in nominal and adverse conditions. Mid and right: application of wind tunnels and CFD to acquire accurate aerodynamic estimates for simulation of flight outside the normal envelope, aircraft damage and icing to mitigate in-flight loss-of-control

7.1 Pilot training

A significant part of LOC-I accidents have been attributed to a lack of the crew's awareness and experience in extreme flight conditions. In the course of loss-of-control events, the aircraft often enters unusual attitudes or other types of upsets (Figure 15). To prevent or timely recover from a loss-of-control or unusual attitude situation, it is essential that the pilot rapidly recognizes the condition, initiate recovery actions and follows appropriate recovery procedures. Inadequate recovery may exacerbate the situation and lead to the loss of the aircraft.

Aviation authorities recognize the need to educate pilots on upset recovery techniques to reduce the amount of LOC-I accidents. As in-flight training with large aircraft is expensive and unsafe, ground-based flight simulators are applied as an alternative to in-flight training of loss-of-control scenarios. Ground-based full flight simulators (FFS) that are capable enough of accurately representing extreme flight conditions would significantly improve the effectiveness of upset recovery training while being part of the standard airline training program.

Fig. 15. Aircraft showing unusual attitude typical during in-flight loss-of-control or upset conditions

Current flight simulators, however, are considered inadequate for the simulation of many upset conditions as the aerodynamic models are only applicable to the normal flight envelope. Upset conditions can take the aircraft outside the normal envelope where aircraft behaviour may change significantly, and the pilot may have to adopt unconventional control strategies (Burks, 2009). Furthermore, standard hexapod-based motion systems are unable to reproduce the high accelerations, angular rates, and sustained G-forces occurring during upsets and the recovery from adverse conditions.

The European Seventh Framework Program Simulation of Upset Recovery in Aviation (SUPRA, 2009-2012) aims to improve the aerodynamic and the motion envelope of ground-based flight simulators required for conducting advanced upset recovery simulation. The research not only involves hexapod-type flight simulators but also experimental centrifuge-based simulators (Figure 16). The aerodynamic modeling within the SUPRA project employs a unique combination of engineering methods, including the application of validated CFD methods and innovative physical modeling to capture the major aerodynamic effects that occur at high angles of attack. The flight simulator motion cueing research within SUPRA aims to extend the envelope of standard FFSs by optimizing the motion cueing software. In addition, the effectiveness of the application of a new-generation centrifuge-based simulations are investigated for the simulation of G-loads that are typically present in upset conditions. Information on the SUPRA program can be found in reference (Groen et al., 2011) and is also available via the website of the project (www.supra.aero).

Fig. 16. SUPRA simulation facilities for conducting advanced upset recovery simulation research to improve pilot training in upset recovery and reduce LOC-I accident rates.
Left: NLR Generic Research Aircraft Cockpit Environment (GRACE). Mid: TsAGI PSPK-102. Right: TNO/AMST Desdemona

8. Summary and conclusion

A benchmark for the integrated evaluation of new fault detection, isolation and reconfigurable control techniques has been developed within the framework of the European GARTEUR Flight Mechanics Action Group FM-AG(16) on Fault Tolerant Control. Validated against data from the Digital Flight Data Recorder (DFDR), the benchmark addresses the need for high-fidelity nonlinear simulation models to improve the prediction of reconfigurable system performance in degraded modes. The GARTEUR RECOVER benchmark is suitable for both offline design and analysis of new fault tolerant flight control system algorithms and integration on simulation platforms for piloted hardware in the loop testing. In conjunction with enhanced graphical tools, including high resolution aircraft visualisations, the benchmark supports tool-based advanced flight control system design and evaluation within research, educational or industrial framework.

The GARTEUR Action Group FM-AG(16) on Fault Tolerant Control has made a significant step forward in terms of bringing novel "intelligent" self-adaptive flight control techniques, originally conceived within the academic and research community, to a higher technology readiness level. The research program demonstrated that the designed fault tolerant control algorithms were successful in recovering control of significantly damaged aircraft.

Within the international aviation community, urgent measures and interventions are being undertaken to reduce the amount of loss-of-control accidents caused by mechanical failures, atmospheric events or pilot disorientation. The application of fault tolerant and reconfigurable control, including aircraft envelope protection, has been recognised as a possible long term option for reducing the impact of flight critical system failures, pilot disorientation following upsets or flight outside the operational boundaries in degraded conditions (e.g. icing). Fault tolerant flight control, and the (experimental) results of this GARTEUR Action Group, may further support these endeavours in providing technology solutions aiding the recovery and safe control of damaged aircraft or in-flight upset conditions. Several organisations within this Action Group currently conduct in-flight loss of control prevention research within the EC Framework 7 program Simulation of Aircraft Upsets in Aviation SUPRA (www.supra.aero). The experience obtained by the partners in this Action Group will be utilised to study future measures in mitigating the problem of in-flight loss-of-control and upset recovery and prevention.

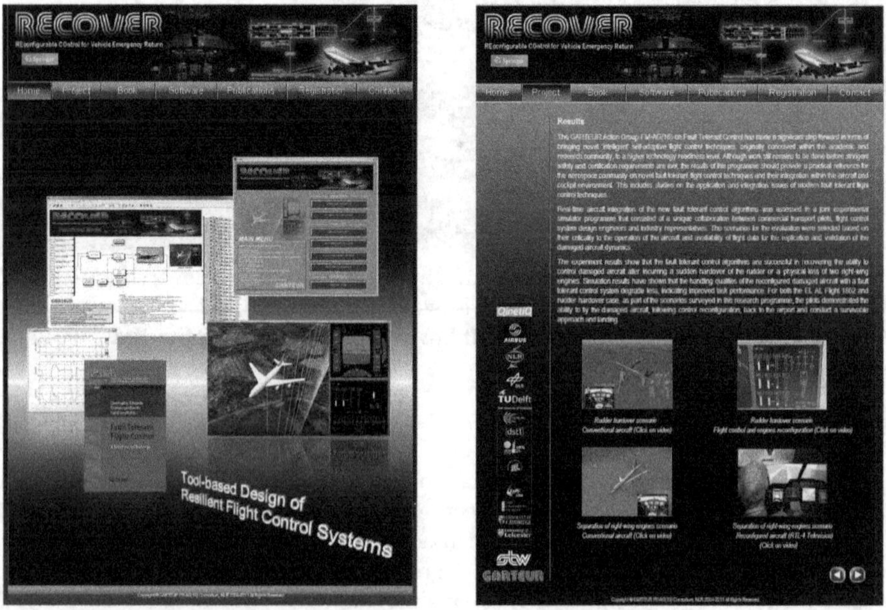

Fig. 17. The GARTEUR FM-AG(16) Fault Tolerant Flight Control project website provides information on the project, links to ongoing research, publications and software registration (www.faulttolerantcontrol.nl)

The results of the GARTEUR research program on fault tolerant flight control, as described in this chapter, have been published in the book 'Fault Tolerant Flight Control - A Benchmark Challenge' by Springer-Verlag (2010) under the Lecture Notes in Control and Information Sciences series (LNCIS-399) (Edwards et al., 2010). The book provides details of the RECOVER benchmark model architecture, mathematical models, modelled fault scenarios and examples for both offline and piloted simulation applications. The GARTEUR RECOVER benchmark simulation model, which accompanies the book, is available via the project's website (www.faulttolerantcontrol.nl) after registration. The website (Figure 17) provides further access to contact information, follow-on projects and future software updates.

9. Acknowledgements

The authors recognise the contributions of the members of the GARTEUR FM-AG(16) Action Group to this chapter. The authors also appreciate the funding that the Dutch Technology Foundation STW has provided as part of the GARTEUR activities. Special thanks to Jaap Groeneweg and Ronald Verhoeven of NLR for their contribution to the RECOVER aircraft visualisation tools. A word of thanks to all those who have contributed to the further improvement of the GARTEUR RECOVER benchmark model within their flight control research programmes, especially Andres Marcos of DEIMOS Space and Gary Balas of the University of Minnesota. The authors would like to thank the SUPRA consortium and especially Eric Groen of the Netherlands Organisation for Applied Scientific Research (TNO) for their contribution to this chapter. The SUPRA project is co-financed by the European Commission under Grant Agreement no. 233543.

10. References

Burcham, F.W. & Fullerton, C.G. (2004). Manual Manipulation of Engine Throttles for Emergency Flight Control, NASA/TM-2004-212045

Burks. B. (2009). Upset Recovery Training: A Call for a Higher Standard of Training, *Royal Aeronautical Society Flight Simulation Conference: Towards the Edge of the Envelope,* Royal Aeronautical Society, London, UK, June 3-4, 2009

Civil Aviation Authority of the Netherlands. (2007). *Civil Aviation Safety Data*

Corder, M. (2004). Crippled But Not Crashed. *Scientific American,* (July 2004)

Edwards, C., Lombaerts, T.J.J. & Smaili, M.H. (Ed(s).). (2010). *Fault Tolerant Control – a Benchmark Challenge,* Lecture Notes in Control and Information Sciences (LNCIS-399), Springer-Verlag, ISBN 978-3-642-11689-6

Ganguli, S., et al. (2005). Piloted Simulation of Fault Detection, Isolation and Reconfiguration Algorithms for a Civil Transport Aircraft, *Proceedings of AIAA Guidance, Navigation and Control Conference,* AIAA-2005-5936, San Francisco, California, 2005

Groen, E., Fucke, L., Goman, M., Biryukov, V. & Smaili, M.H. (2011). Improving Flight Safety by Pushing the Training Envelope: The European Project SUPRA, *Proceedings of the 3rd CEAS Air & Space Conference / 21st AIDAA Congress,* Venice, Italy, October 24-28, 2011

Lombaerts, T.J.J. et al. (2006). Assessment Criteria as Specifications for Reconfiguring Control, *Proceedings of AIAA Guidance, Navigation and Control Conference,* Keystone, Colorado, USA, August, 2006

Lombaerts, T.J.J., Chu, P., Mulder, J.A. & Joosten, D.A. (2011). Modular Flight Control Reconfiguration Design and Simulation. *Control Engineering Practice, Elsevier,* Vol. 19, No. 6, (2011), pp. 540–554, DOI: 10.1016/j.conengprac.2010.12.008

Marcos, A., Balas,G.J. (2001). Linear Parameter Varying Modeling of the Boeing 747-100/200 Longitudinal Motion, *Proceedings of AIAA Guidance, Navigation and Control Conference,* AIAA-2001-4347, Montreal, Canada, 2001

NASA. (2011). Aviation Safety Program, In: *Aviation Safety Program Fact Sheet (NF-2011-04-535-HQ),* 09.09.2011, Available from http://www.aeronautics.nasa.gov/programs_avsafe.htm

Netherlands Aviation Safety Board. (1994). *EL AL Flight 1862, Aircraft Accident Report 92-11,* Hoofddorp

Patton, R.J. (1997). Fault-Tolerant Control Systems: The 1997 Situation, *Proceedings of the IFAC Symposium on SAFEPROCESS,* pp. 1033-1055, Hull, UK, August, 1997

Stroosma, O., Smaili, M.H. & Mulder, J.A. (2009). Pilot-in-the-Loop Evaluation of Fault Tolerant Flight Control Systems, *Proceedings of the IFAC SAFEPROCESS 2009 Conference,* Barcelona, Spain, 2009

Smaili, M.H. (1997). *Flight Data Reconstruction and Simulation of EL AL Flight 1862,* Graduation Report, Delft University of Technology, Delft

Smaili, M.H. & Mulder, J.A. (2000). Flight Data Reconstruction and Simulation of the 1992 Amsterdam Bijlmermeer Airplane Accident, *Proceedings of AIAA Modeling and Simulation Conference,* AIAA-2000-4586, Denver, Colorado, August, 2000

The Boeing Company. (1999). Intelligent Flight Control: Advanced Concept Program – Final Report, BOEING-STL 99P0040

Wright Laboratory. (1991). *Self-Repairing Flight Control System,* Final Report, WL-TR-91-3025

Zhang, J. & Jiang, J. (2003). Bibliographical Review on Reconfigurable Fault-Tolerant Control Systems, *Proceedings 5th IFAC Symposium on Fault Detection, Supervision and Safety for Technical Processes 2003,* Washington, D.C., USA, June 9-11, 2003

Zhang, Y. (2005). Fault Tolerant Control Systems: Historical Review and Current Research, *Presented at the Centre de Recherche en Automatique de Nancy,* Universite Henri Poincare, Nancy 1, France, December 5, 2005

Permissions

The contributors of this book come from diverse backgrounds, making this book a truly international effort. This book will bring forth new frontiers with its revolutionizing research information and detailed analysis of the nascent developments around the world.

We would like to thank Dr Ir Thomas Lombaerts, for lending his expertise to make the book truly unique. He has played a crucial role in the development of this book. Without his invaluable contribution this book wouldn't have been possible. He has made vital efforts to compile up to date information on the varied aspects of this subject to make this book a valuable addition to the collection of many professionals and students.

This book was conceptualized with the vision of imparting up-to-date information and advanced data in this field. To ensure the same, a matchless editorial board was set up. Every individual on the board went through rigorous rounds of assessment to prove their worth. After which they invested a large part of their time researching and compiling the most relevant data for our readers. Conferences and sessions were held from time to time between the editorial board and the contributing authors to present the data in the most comprehensible form. The editorial team has worked tirelessly to provide valuable and valid information to help people across the globe.

Every chapter published in this book has been scrutinized by our experts. Their significance has been extensively debated. The topics covered herein carry significant findings which will fuel the growth of the discipline. They may even be implemented as practical applications or may be referred to as a beginning point for another development. Chapters in this book were first published by InTech; hereby published with permission under the Creative Commons Attribution License or equivalent.

The editorial board has been involved in producing this book since its inception. They have spent rigorous hours researching and exploring the diverse topics which have resulted in the successful publishing of this book. They have passed on their knowledge of decades through this book. To expedite this challenging task, the publisher supported the team at every step. A small team of assistant editors was also appointed to further simplify the editing procedure and attain best results for the readers.

Our editorial team has been hand-picked from every corner of the world. Their multi-ethnicity adds dynamic inputs to the discussions which result in innovative outcomes. These outcomes are then further discussed with the researchers and contributors who give their valuable feedback and opinion regarding the same. The feedback is then collaborated with the researches and they are edited in a comprehensive manner to aid the understanding of the subject.

Apart from the editorial board, the designing team has also invested a significant amount of their time in understanding the subject and creating the most relevant covers. They scrutinized every image to scout for the most suitable representation of the subject and create an appropriate cover for the book.

The publishing team has been involved in this book since its early stages. They were actively engaged in every process, be it collecting the data, connecting with the contributors or procuring relevant information. The team has been an ardent support to the editorial, designing and production team. Their endless efforts to recruit the best for this project, has resulted in the accomplishment of this book. They are a veteran in the field of academics and their pool of knowledge is as vast as their experience in printing. Their expertise and guidance has proved useful at every step. Their uncompromising quality standards have made this book an exceptional effort. Their encouragement from time to time has been an inspiration for everyone.

The publisher and the editorial board hope that this book will prove to be a valuable piece of knowledge for researchers, students, practitioners and scholars across the globe.

List of Contributors

Ahmed Mohamed and Apostolos Mamatas
University of Florida, USA

Urbano Tancredi
University of Naples Parthenope, Italy

Federico Corraro
Italian Aerospace Research Centre, Italy

Xiaojun Xing and Dongli Yuan
Northwestern Polytechnical University, Xi'an, China

Youmin Zhang and Abbas Chamseddine
Department of Mechanical and Industrial Engineering, Concordia University, Canada

Jie Zeng
ZONA Technology, Inc., USA

Raymond De Callafon
University of California, San Diego, USA

Martin J. Brenner
NASA Dryden Flight Research Center, USA

Marilena D. Pavel
Faculty of Aerospace Engineering, Delft University of Technology, The Netherlands

Hafid Smaili
National Aerospace Laboratory NLR/Delft University of Technology, NLR Cockpit & Flight Operations Department, Amsterdam, The Netherlands

Jan Breeman
National Aerospace Laboratory NLR/Delft University of Technology, NLR Avionics & Systems Department, Amsterdam, The Netherlands

Thomas Lombaerts
DLR Institute of Robotics and Mechatronics, Wessling, Germany

Printed in the USA
CPSIA information can be obtained
at www.ICGtesting.com
JSHW011406221024
72173JS00003B/442

9 781632 400734